KV-587-502

Laboratory Work in Soil Mechanics

Second Edition

Brian Vickers

B.Sc.(Hon), M.Sc., C.Eng., M.I.C.E.
Head of Department of Civil Engineering
Bolton Institute of Higher Education
(incorporating Bolton Institute of Technology)

GRANADA
London Toronto Sydney New York

Granada Publishing Limited – Technical Books Division
Frogmore, St Albans, Herts AL2 2NF
and
36 Golden Square, London W1R 4AH
515 Madison Avenue, New York, NY 10022, USA
117 York Street, Sydney, NSW 2000, Australia
100 Skyway Avenue, Rexdale, Ontario, Canada M9W 3A6
61 Beach Road, Auckland, New Zealand

Copyright © 1978, 1983 by Brian Vickers

British Library Cataloguing in Publication Data
Vickers, Brian
  Laboratory work in soil mechanics.—2nd ed.
  1. Soils—Testing—Laboratory manuals
  I. Title
  624.1′5136′0287   TA710.5

  ISBN 0–246–11819–9

First published in Great Britain 1978 by Crosby Lockwood Staples
(ISBN 0 258 97084 7) under the title Laboratory Work in Civil
Engineering: Soil Mechanics
Reprinted 1979
Second edition 1983

Printed in Great Britain by Fletcher & Son Ltd, Norwich

All rights reserved. No part of this publication may be reproduced,
stored in a retrieval system, or transmitted in any form or by any
means, electronic, mechanical, photocopying, recording or otherwise,
without the prior permission of the publisher.

Granada ®
Granada Publishing ®

Contents

Preface to Second Edition

Initially I should like to thank those people who have contacted me since the First Edition was published in 1978. I have taken note of all the comments received and I have tried to incorporate most of the suggested changes in this revised edition. I want to thank particularly T. A. H. Dodd, of the University of Canterbury, New Zealand, and E. N. Bromhead, of Kingston Polytechnic, for especially helpful comments.

Essentially the text is intended for student use and I continue to hope that it will form the basis for further study and related experience.

Mrs Mildred Jones is sincerely thanked for her devotion in typing the revised manuscript.

Brian Vickers

Bolton

Preface to First Edition

The aim of this book is to meet the needs of students undertaking undergraduate and technician courses in civil engineering, by presenting a concise, clear introduction to laboratory testing techniques, procedures and limitations and thus providing a sound basis for further study and related experience.

Each chapter briefly outlines the appropriate soil mechanics theory and the practical purposes of the various types of test. This is followed by detailed test procedure and an explanation of the equipment used. Each test is summarised, with comments on its benefits and disadvantages, and it is hoped that these summaries will encourage further reading, thought and discussion. Datasheets in manuscript form, including graphs and calculations, are presented for many of the experiments described.

Readers are recommended to purchase a copy of BS 1377: 1975, *Methods of test for soil for civil engineering purposes*, published by the British Standards Institution and obtainable from the BSI at 2 Park Street, London W1A 2BS.

The author wishes to express his sincere gratitude to Mr Stephen A. Matthews, who prepared the diagrams and datasheets, and to Mrs Mildred Jones, who typed the manuscript.

Brian Vickers

Bolton

CHAPTER 1

Introduction

1.1 The Purpose of Laboratory Testing of Soils

A wide variety of problems associated with soil mechanics and foundation engineering are solved by a combination of theoretical knowledge, of knowledge of precedents, of knowledge of the geology and history of the site under consideration, of knowledge of the properties of the soil under consideration obtained from laboratory and field tests, and of an appreciation of the function of the proposed structure and therefore of its most important features. These features may include those that affect settlement predictions, strength requirements, stability, effects of ground water and consideration of the effects of time upon these factors.

In some cases, for preliminary calculation purposes or for outline design proposals, some of the essential information can be approximated using reliable data from similar projects undertaken in similar soil conditions. More often, however, the soil properties and subsequent engineering judgements are based upon laboratory tests performed on representative samples taken, during a site investigation, from the site of the proposed works. The scope, type and quality of the corresponding site investigation is important and is decided upon only after considering the subsequent laboratory test programme.

The investigation and test programme are interrelated and, in fact, it may well be that a pilot site investigation, involving sampling, is carried out to establish the type and characteristics of the soil to be studied. From this pilot investigation a laboratory test programme is then decided upon, which needs to take into account the size of samples, the quality of samples and the frequency of sampling with respect to both lateral and vertical variations in the material.

Obviously the potential problems of the proposed structure or construction must also be studied simultaneously since they affect aspects of both the investigation and the test programme.

Having decided on the appropriate laboratory tests, the major site investigation is undertaken with the distinct purpose of fulfilling a considered test programme. It is important that this major investigation incorporates a degree of flexibility since, as more is revealed about the site under consideration, it may be necessary to reconsider, for instance, the number of samples to be taken. The result may well be an increase or a decrease in the proposed period, scope and quantity of laboratory testing. The overriding question should be: 'How best will the correct results be obtained for this particular project to be undertaken at this particular site?' Everything possible should then be done to achieve this state of perfection.

It is next the responsibility of the engineer in charge to present a report based upon the results and findings of the investigation and test programme. This, again, is an extremely important part of the procedure and the contents of the report should be relevant, sensible and concise. It usually includes recommendations about various aspects of the project such as construction procedures, instrumentation and necessary theoretical analysis, together with a summary of the test results upon which such recommendations are based.

From the foregoing it is obvious that the complete construction process, from conception of a scheme to completed construction and subsequent satisfactory use, is complex and depends upon reasoned judgement at all stages. This book is concerned with one aspect, namely soil laboratory testing, which constitutes an important element of the complete integrated process.

1.2 Soil Identification

Before any testing and subsequent analysis is undertaken it is essential to identify and describe adequately the soil samples; it is the responsibility of the laboratory worker to make notes, generally classify and describe all such samples from an initial examination. Such a description may only duplicate that of the driller in charge of the site investigation rig but, nevertheless, it serves as a vital check on information supplied from site. This information includes details such as sample numbering, sample depth and so on. A further benefit of laboratory description is that any subsequent test results can be related to it as an aid to the final evaluation and judgement of the results.

The description and identification must include features of the soil as visually observed (e.g. fissures, the presence of layering, organic rootlets, sand layers, the presence of boulders, etc.). In addition it is necessary to observe the physical 'feel' of the material, i.e. when worked in the hand, subjected to pressure by the thumb, finger-nail, etc., and to note any distinctive features (Table 1.1). The accuracy of the description assigned to a soil depends largely upon the experience of the laboratory technician, in performing these simple tests and drawing appropriate conclusions. Consequently it is of fundamental importance that students and trainees acquire

Table 1.1 Simple Soil Identification Tests and Observations

Gravels and gravelly soils	Easily identified by visual inspection. Absence of dry strength indicates zero clay content.
Sands and sandy soils	Most particles visible to naked eye. Feel gritty when rubbed between fingers. Absence of dry strength indicates zero clay content.
Silts	Not gritty. Cannot easily be rolled into threads when moist.
Clays: medium plasticity ($LL = 35\%-50\%$)	Easily rolled into threads. Shrink on drying.
Clays: highly plastic ($LL > 50\%$)	Easily rolled into threads. Shrink considerably on drying. Feel greasy.

quickly the necessary techniques: alert observation of soil characteristics at this stage of laboratory testing provides both an indication of the problems to be encountered and guidance as to the probable behaviour of the soil when being tested.

1.3 Soil Samples

Most present-day laboratory tests employed in soil mechanics are highly developed and perfected. It is necessary to recognise the relation between samples and subsequent testing. Much skill, experience and time may have been used by the site investigation team in obtaining the samples from the site under consideration. It is therefore important that the method of sampling, exact location with respect to plan and elevation, date of sampling and all other relevant information are correctly recorded in the laboratory. As far as the samples are concerned, the main requirement is that they are representative of the mass of the strata from which they have been taken. This involves decisions about the size of sample, the method of sampling and the location of sampling.

Undisturbed samples can be obtained either by employing some type of sampler, usually incorporating the use of a sample tube, or by taking the sample from the face of an excavation such as a trial pit, and immediately covering it with a protective, impervious layer of wax.

Ideally, samples should be tested within a short time of arrival at the laboratory. This is because it has become evident that satisfactory storage of soil samples, maintaining natural moisture content and other properties, is difficult. In addition, early results obtained from testing the initial samples received from a site may well indicate that more samples or larger samples need to be taken, so that a revised programme and procedure for the sampling becomes necessary. Inevitably some storage is needed and may even be essential if further studies are to be undertaken, consequently facilities for storage should be adequate in terms of space and of temperature and humidity control.

Both types of undisturbed sample (from tubes or from pit excavations) are sealed by wax placed immediately after sampling. The choice of wax is important and such factors as shrinkage, strength, brittleness and permeability must be considered. The wax seal usually incorporates a layer of cheesecloth and the thickness of coating can vary from 12 mm to 25 mm, depending upon the function of the seal. Generally the wax has a hydrocarbon base; paraffin wax is commonly used, although it has been shown that its performance can be somewhat inferior to that of other waxes.

Soil stored in metal tubes can be seriously affected by the material from which the tube is manufactured. Some electrolytic and chemical action between soil and metal can take place, with serious consequences if the storage is long-term. Sample tubes can be treated by plating to minimise this action; alternatively, PVC sample tubes can be employed, provided that they maintain their original shape and do not distort. Obviously cleanliness is important and the use of grease and oils internally may minimise the undesirable chemical action.

Fundamental to good laboratory work are the seemingly simple tasks of labelling, noting the depth of sampling and recognising its orientation and sequence. Here the use of waxed marking instruments is essential. In handling the samples when extruded, the bare hands should never be in contact with the undisturbed sample as this can lead to loss of moisture and possible further disturbance. Similarly, proper support of the sample is essential to prevent fracture or cracking.

Cutting and trimming should be performed carefully, quickly and accurately, and the use of mitre boxes and jigs can be advantageous. The cutting tools chosen should be appropriate to the material being handled and may include cheese wires, sharp knives and even saw blades; these implements should be available in a variety of sizes. The most important consideration is to prevent distortion, disturbance and smear of potential drainage faces. Any physical motion applied to a soil sample must be steady, slow and considered rather than haphazard and jerky.

In some cases the 'splitting' of samples can prove a useful technique when identification of the structure and type of soil is required. This can be performed on, say, 100 mm diameter samples with a minimum of smear. If the split sample is photographed immediately after splitting and subsequently after periods of drying, useful information can be obtained with respect to layering and permeability.

This technique of 'splitting the samples' is invaluable in aiding soil description, in assessing potential sampling and testing problems and in identifying any minor details which may affect the likely test and full-scale behaviour of the soil. Such details can include

layering,   fissuring   and   intrusions.

Presence of any granular material can be observed and note taken of the size and shape of such particles.

Thus one could describe, for example, a clay as stiff, reddish-brown boulder clay with small, rounded boulders (10 mm size approx.) with evidence of some organic rootlets.

The soil sample must be split naturally so that any 'smear' effects are eliminated or minimised. This can be achieved by inserting purpose-made 'splitters' along one generator of the cylindrical sample. These 'splitters' run the full length of the sample and are inserted for about 20% of the sample diameter, as shown in Fig. 1.1. Pocket penetrometers can then be inserted

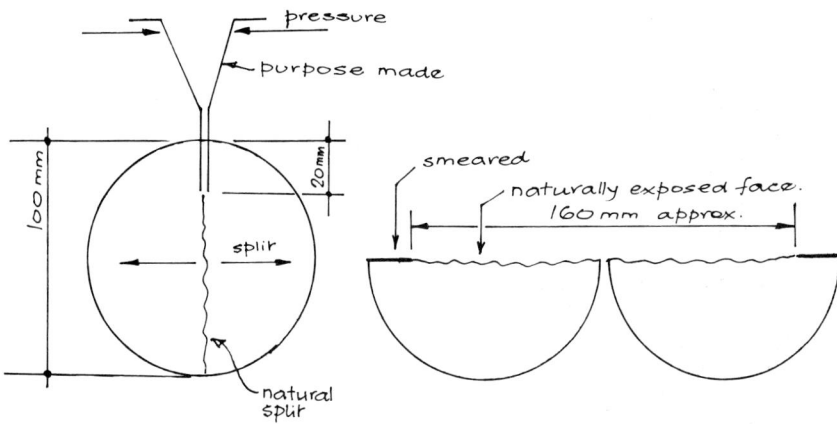

Fig. 1.1

along the 'smeared' exposed edges of the sample to give some comparison to relative stiffness descriptions of the soil.

It has also been found useful when extruding such samples, for splitting, to extrude them into a 100 mm PVC or similar pipe, split along two opposing generators. This then ensures that handling and extrusion are carefully managed. Experience has shown that wide rubber bands, used to hold the two halves of the PVC tube during extrusion, are helpful. This allows the 100 mm sample to be extruded into a cylinder of slightly larger diameter thus, once again, minimising any drag or disturbance effects. This technique has been undertaken very easily with sands, even fine-grained ones, and here the PVC tubes help to retain the split sample for photographic purposes and prevent complete collapse of the sample as it dries out.

If a borehole is continuously sampled in this way, the samples can be laid end to end on the laboratory floor thus giving a complete and interesting record of the soil profile. It is important to retain all the samples in their *in situ* condition, including the samples recovered in the detachable cutting shoe of the sampler. If these latter, vital parts of the sample are not retained then there could be serious gaps within the complete borehole log information.

A series of such samples, taken recently from a coastal site, are included in Fig. 1.2.

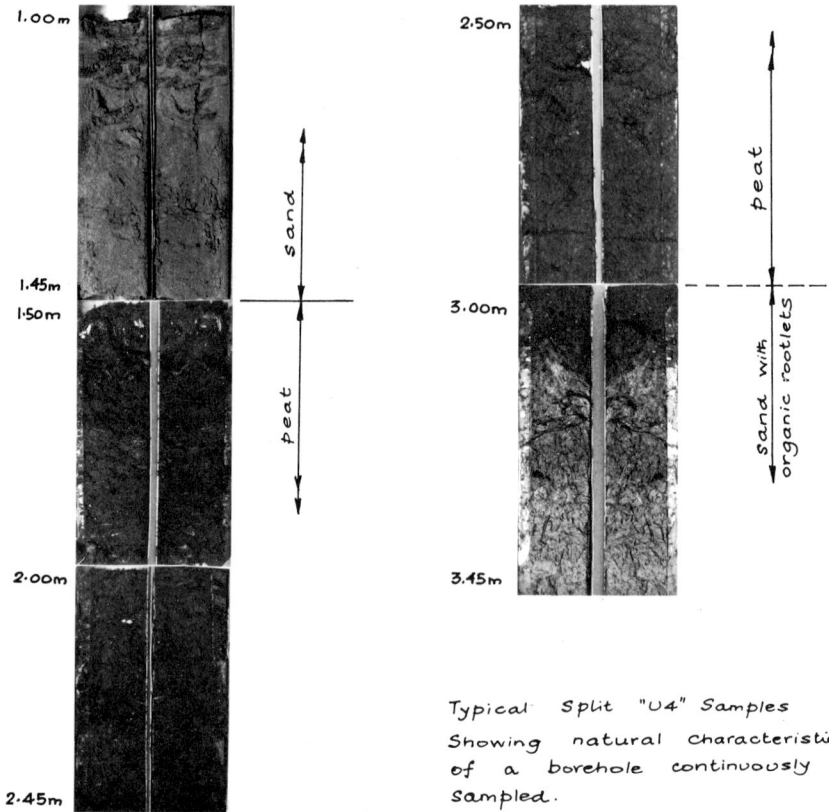

Fig. 1.2

It can be observed that from 1·00 m down to 1·45 m the sand changes colour from light brown to dark brown and that there is little evidence of any large particles, perhaps indicating a wind-blown deposit. From 1·5 m there is a layer of fibrous peat, dark-reddish brown with evidence of silver birch rootlets. The interface of the peat and the sand below the peat, at just over 3·0 m depth is of interest. It clearly shows that rootlets are present in the lower sand and that these are of the reed-type rather than cylindrical rootlets. Such organic intrusions would be potential drainage paths. In addition the photographs illustrate the variability of soil and the need, at this site, to test samples that fully portray the soil mass.

The smear areas can be seen at each side of the photographs and the split retaining tubes can clearly be seen.

The retention of such photographs is a clear aid to any further discussions or reports made about the particular site and will overcome many of the personal variations in describing colour, texture or even soil type.

1.4 Determination of Unit Weight and Moisture Content

Determination of the unit weight of a soil and of its natural moisture content are perhaps the soil tests most commonly undertaken. With experience, they can provide excellent indications of the state of the material and of its characteristic strength and other properties.

Unit Weight
The determination of the unit weight of a soil is a comparatively simple operation for undisturbed soil samples. Where the volume of a particular sample tube is known, the tube is weighed empty and when full of soil, so that the bulk density ρ is given by

$$\rho = \frac{M_2 - M_1}{V} = \frac{\text{total mass}}{\text{total volume}}$$

where M_1 = mass of empty sample tube
M_2 = mass of sample tube plus soil
V = volume of sample tube.

Convenient units for the bulk density ρ are kg/m^3 or Mg/m^3: thus if M_1 and M_2 are observed in grammes, the volume must be determined in cubic centimetres.

Since

$$1 \text{ g/cm}^3 = 1000 \text{ kg/m}^3$$

therefore

$$\rho = \frac{M_2 - M_1}{V} \times 1000 \text{ kg/m}^3 = \frac{M_2 - M_1}{V} \text{ Mg/m}^3$$

Obviously accuracy of weighing and the determination of dimensions for volume measurement are important. Results should be reported to the nearest 1 kg/m^3.

The unit weight γ of a soil is defined as the ratio of the total weight (a force) to total volume, that is

$$\gamma = \frac{M \times g}{V} = \frac{w}{V}$$

where $g = 9.81 \text{ m/s}^2$ = acceleration due to gravity, and w = weight of soil (N).

Thus

$$\gamma = \rho \cdot g \ \mathrm{N/m^3}$$

or, more conveniently, the bulk unit weight $\gamma = \rho \cdot g/1000 \ \mathrm{kN/m^3}$, with ρ expressed as $\mathrm{kg/m^3}$.

Sometimes g is approximated to a value of 10 rather than 9·81, giving the bulk unit weight

$$\gamma = \frac{\rho \cdot 10}{1000} = \frac{\rho}{100} \ \mathrm{kN/m^3}$$

where ρ is the bulk density.

Moisture Content

The moisture content of soil is determined as follows.

Depending upon the soil type, the following quantities are taken as being sufficiently representative samples:

Fine-grained soils	30 g
Medium-grained soils	500 g
Coarse-grained soils	3000 g = 3 kg

A suitable container with lid needs to be available for each sample. First, the container is weighed (M_1), complete with lid, after ensuring that it is absolutely dry and clean. The appropriate weight of wet soil is then placed in the container in a loose state. The lid is securely fixed and the container, complete with lid and wet soil, is then weighed again (M_2). The lid is then removed, placed in an oven together with the container and contents and dried at a temperature of 105° to 110° C, usually for a period of 24 hours.

It is important to note that some soils may be susceptible to fundamental change if they are dried at a temperature of 105° to 110° C; examples are organic soils and soils with a significant gypsum content. With such soils lower drying temperatures must be used and the period of drying may have to be prolonged. A record of any such changes must be noted on the corresponding results sheet.

The drying time of 24 hours may be reduced when it is proven that weights recorded at four hour intervals do not differ by more than 0·1% of the original mass of the wet sample.

When the drying period is complete, the lid is replaced and weighed with the container and any soil (M_3), after having been allowed to cool naturally.

The moisture content is then calculated using the definition of moisture content:

$$m = \frac{\text{mass of water}}{\text{mass of solid matter}} \times 100\%$$

that is

$$m = \frac{M_2 - M_3}{M_3 - M_1} \times 100\%$$

where m = moisture content (sometimes referred to as the water
content and given the symbol w)

M_1 = mass of container + lid
M_2 = mass of container + lid + wet soil
M_3 = mass of container + lid + dry soil.

The important points to be observed concerning moisture content determination are:

1. The sample must be chosen so that it is a representative sample of the soil under consideration.
2. All risk of drying during handling and weighing must be minimised.
3. The sample must be sufficiently large to give adequate accuracy of weighing and the choice of balance must be appropriate to the size of the sample.
4. The container must be thoroughly cleansed and dried before use. It must be manufactured or treated so that it is non-corrodible.
5. The weighing of containers and contents which are above room temperature is not recommended because there is an increased risk of dropping the container and the accuracy of the balance can appreciably be affected by temperature. The use of a desiccator can prove useful in cooling plastic soils which may otherwise take more than 2 hours or so to cool.
6. It must be noted that the term 'dry' soil is not strictly accurate. Soil is usually defined as being 'dry' when it has attained constant weight after being heated to 105° to 110° C, but there is undoubtedly still some water present in the soil in the form of adsorbed water surrounding the soil solid particles. Much higher temperatures are required to drive off this water.

1.5 Relationships Between the Various Phases of a Soil

The various relationships between the three phases of a soil, namely air, water and solid matter, are listed below. Taking the model soil sample as shown in Fig. 1.3, each relationship is defined in terms of the symbols and some important notes are added.

$$\text{Moisture content} = m = \frac{\text{mass of water}}{\text{mass of solid}} = \frac{M_w}{M_s} \text{ (usually \%)}$$

$$\text{Degree of saturation} = S = \frac{\text{volume of water}}{\text{volume of voids}} = \frac{V_w}{V_v} = \frac{V_w}{V_a + V_w}$$

Note: 1. m can be greater than 100%.
2. If the soil is fully saturated (i.e. no air present), $S = 100\%$; if the soil

is dry, $S = 0\%$. A partially saturated soil has S of intermediate value (e.g. $S = 78\%$).

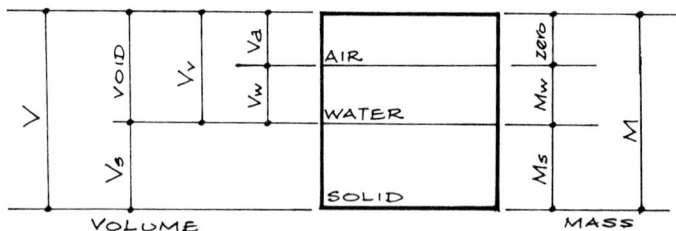

Fig. 1.3

$$\text{Void ratio} = e = \frac{\text{volume of void}}{\text{volume of solid}} = \frac{V_v}{V_s} = \frac{V_a + V_w}{V_s}$$

Note: e can be greater than unity.

$$\text{Porosity} = n = \frac{\text{volume of void}}{\text{total volume of soil}} = \frac{V_v}{V} = \frac{V_a + V_w}{V_a + V_w + V_s}$$

Also

$$n = \frac{e}{1 + e}$$

$$\text{Specific gravity of the solid matter} = G_s = \frac{\text{mass of solid}}{\begin{array}{c}\text{mass of water occupying}\\\text{the same volume as the}\\\text{solid matter in the soil}\end{array}} = \frac{M_s}{V_s \rho_w}$$

where ρ_w = density of water at $20°$ C = $1000 \text{ kg/m}^3 = 1 \text{ Mg/m}^3$.

Note: G_s is usually in the range 2·5 to 2·8 for most soils found in the United Kingdom. A method for its determination is given in Section 2.4.

$$\text{Bulk density} = \rho = \frac{\text{total mass of soil}}{\text{total volume of soil}} = \frac{M_w + M_s}{V_a + V_w + V_s} = \frac{M}{V}$$

It can be shown that

$$S \cdot e = \frac{m}{100} G_s$$

and, for the special case when $S = 100\%$ (fully saturated soil),

$$e = \frac{m}{100} G_s$$

It can also be shown that, where the volume of solid matter is unity (i.e. $V_s = 1$),

$$\text{Bulk density} = \rho = \rho_w \frac{G_s + (S/100)e}{1 + e}$$

where e = volume of void if $V_s = 1$, since $e = V_v/V_s = V_v/1$.

Note: If S = 0% (dry soil), $\rho = \rho_d = \rho_w G_s/(1 + e)$ = dry density.
   If $S = 100\%$ (fully saturated soil), $\rho = \rho_{sat} = \rho_w(G_s + e)/(1 + e)$
   = saturated density.

The bulk unit weight γ of a soil is defined as the ratio of total weight to the total volume (note that the term 'weight' signifies a force).

Thus   Weight = mass × acceleration

or

$$w = M \times g$$

where g = acceleration due to gravity = 9·81 m/s².

Therefore

$$\text{Bulk unit weight} = \gamma = w/V = M \times g/V(kN/m^3)$$

or $$\gamma = \frac{G_s + (S/100)e}{1 + e}\gamma_w$$

where γ_w = bulk unit weight of water = 9·81 kN/m³.

1.6 Laboratory Test Reports

Commercial or other soils laboratory test reports play a vital part in any civil engineering project.

   The form of a test report varies according to its function. The report is an important part of the communication process and must be clear, well-reasoned and concise. Note should be made of the test procedures adopted, and the results should be presented in both tabular and graphical form. Background information to the initiation of the testing is essential and a summary of the geological aspects of the site, with special reference to unusual characteristics, is useful. Particulars of the site investigation, with respect to type of sampling, frequency of sampling, etc., should be given. Reasoned arguments are essential to justify the procedures adopted.

   The report may also include some analysis based upon the laboratory tests and interpretation of the results of the test programme. Often certain recommendations to the client are then presented, these being the result of discussions, thought and consideration of the whole spectrum of investigation, laboratory testing, analysis, experience and judgement, treated as related and interdependent activities.

For the student, it is essential to become familiar with laboratory test equipment and to take every opportunity to relate theory to practice and gain a wide experience of experimental work. Procedures, equipment, test results and difficulties encountered should all be considered and observed in a critical spirit. The drafting and presentation of formal laboratory test reports is then not only a useful professional exercise, but also a valuable method of clarifying and consolidating laboratory experience.

Most soil mechanics laboratory reports have specific aims and objectives which must head the report. This can be followed by a description of the apparatus, aided by well-proportioned sketches, describing the function of each item of equipment. If equipment and procedure are standard, a reference to a certain standard may suffice. A description of the test procedure followed during the test is essential, and particular note should be made of any special techniques employed. The results are then presented in an orderly way and the appropriate graphs are drawn. Sample calculations are presented in this section and the source is given of any equations or formulae which have been employed.

The discussion of the results and the consideration of how well the aims and objectives of the test have been met usually forms a most important part of the report. This section must include items such as possible sources of error, difficulties experienced, reasons or suggested reasons for unexpected or unusual results, comments on the accuracy of observations, suggestions for modifications to the equipment and amendments to the procedure. Results obtained during testing must be compared with published results for similar soils. It is also useful to discuss possible sources of error that may arise when the laboratory results are used to predict full-scale field behaviour.

It is good practice to provide an index to the report and to establish a cross-reference system so that the reader can easily refer to the relevant description or graph when considering the arguments presented. Where many results and graphs are included, it may be convenient to present these as an appendix to the main report.

CHAPTER 2

Soil Classification Tests

Determination of liquid limit and plastic limit
Particle-size analysis
Specific gravity of solid particles

2.1 Introduction

The size of particles within a soil mass can vary from less than 0·001 mm to over 100 mm. The British Standard size-ranges are given in Table 2.1.

Table 2.1   Particle size (mm) plotted to a logarithmic scale

Clay	Silt			Sand			Gravel			Cobbles and Boulders
	fine	medium	coarse	fine	medium	coarse	fine	medium	coarse	

0·002 0·006   0·02   0·06   0·2   0·6   2   6   20   60 mm

The term 'particle' refers to an individual mineral grain within the soil mass. From the sizes given in Table 2.1 it can be seen that each of the words 'clay', 'silt', 'sand', etc., refers to a range of particle or grain sizes. In some ways it is unfortunate that the same words are used to describe particular types of soil, since soils are normally composed of particles from two or more particle size-ranges. Thus a soil may be described as a 'sandy clay' – a description that implies that the soil has significant proportions of particles from both the sand size-range and the clay size-range; usually the size-range containing the highest percentage of particles is the last-named range in the description.

Coarse-grained soils are usually composed of particles from the sand and gravel size-ranges, while fine-grained soils are made up of particles from the silt and clay size-ranges (i.e. less than 0·06 mm).

The main purpose of the classification system is to provide a description of a soil that will be understood by other engineers. Experience has shown that within each class of soil similar characteristics are displayed. Generally, with coarse-grained soils these similar characteristics are dependent upon the size of the particles and the way in which these sizes are distributed within the soil mass, while with fine-grained soils the moisture content of the soil and the clay mineral particles play important roles in the behaviour of the soil.

Depending upon the moisture content of a fine-grained soil, it can exist in a liquid, plastic, semi-solid or solid state (Fig. 2.1). The change from one state to another (e.g. from liquid to plastic) is (a) gradual and not clearly defined and (b) different for each soil. A decrease in moisture content of the soil results in a decrease in volume; the general relationship between volume and moisture content is shown in Fig. 2.1.

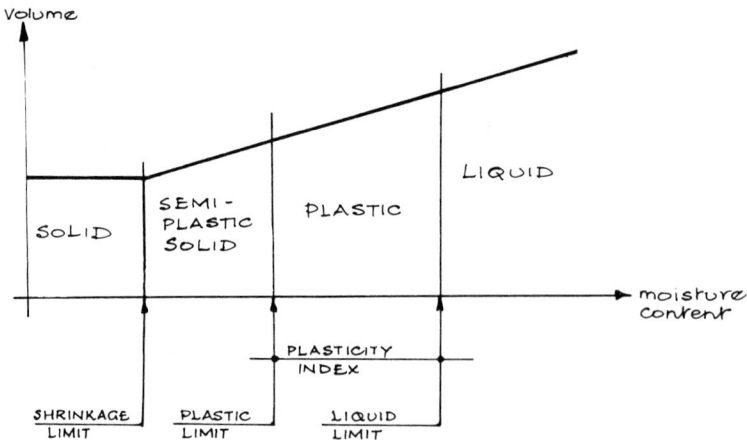

Fig. 2.1 Relationship between soil volume and moisture content

As the transition from state to state is gradual, the points of transition are arbitrarily defined by limits, which are expressed in terms of moisture content. The two most important limits are the liquid limit (LL) and the plastic limit (PL), which define the upper and lower bounds of the plastic state. The difference between these limits is the plasticity index (PI) of the soil.

Other useful terms often used in engineering reports on soil conditions include:

(a) *liquidity index* (LI), which relates the natural moisture content of the soil to both the plastic and liquid limits by the relationship

$$LI = \text{liquidity index} = \frac{m - PL}{PI}$$

where m = the natural moisture content of the soil;

(b) *shrinkage limit*, which defines arbitrarily the change of state from a semi-solid to a solid state and is the moisture content at which the soil remains at a constant volume on drying out;

(c) *activity*, which is the ratio of the plasticity index to the percentage of particles in the clay size-range (i.e. <0.002 mm). Activity is a measure of the degree of plasticity of the clay-size particles.

2.2 Determination of Liquid and Plastic Limits

2.2.1 Liquid Limit Tests

Two methods are currently employed in commercial soils-testing laboratories for the determination of the liquid limit of fine-grained soils. These are based upon using either the Casagrande apparatus or the cone penetrometer. In both methods the sample preparation is the same.

The sample is dried sufficiently to allow it to be broken up by pestle and mortar and care is taken not to crush individual particles. The material used should pass the 425 μm BS test sieve and a minimum mass of 200 g is required for testing.

The dried soil is then thoroughly mixed with distilled water on a clean flat glass plate. A palette knife is used for mixing and enough distilled water is added to form a thick paste. The paste is then placed in an airtight container for 24 hours so that the added water can thoroughly permeate the soil, resulting in a thick homogeneous paste.

Method Using Casagrande Apparatus

The apparatus is shown in Fig. 2.2. Fundamentally it consists of a standard brass cup resting on a hard rubber base. The cup is pivoted and a cam mechanism allows the cup to be repeatedly raised and dropped 10 mm. The cam is operated by a handle and a rev. counter records the number of blows imparted to the brass cup as it is dropped on to the rubber base.

After 24 hours in an airtight container the soil sample is removed and remixed for at least 10 minutes. Some of the thick remixed homogeneous paste is then placed in the brass cup, resting in the lower position. The soil paste is scribed so that its surface is parallel to the rubber base. A standard

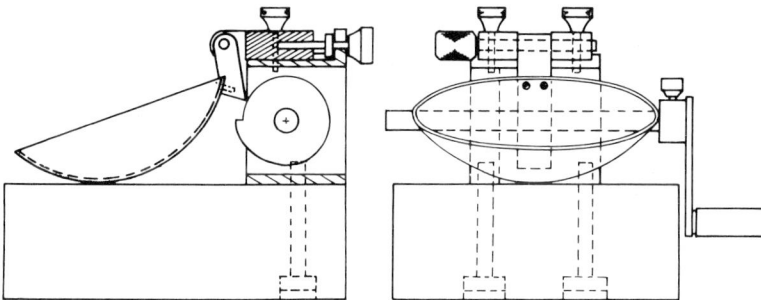

Fig. 2.2 Casagrande liquid limit apparatus

grooving tool (Fig. 2.3a) is then drawn through the paste along a diameter through the pivot so that a section of the cup is as shown in Fig. 2.3b. Care must be taken when forming the groove to ensure that the grooving tool is always held normal to the cup (i.e. along a radius) and that the chamfered edge of the tool leads the formation of the groove.

The handle is then turned at two revolutions per second, lifting and dropping the cup until two parts of the soil close the groove for a distance of 13 mm. The number of blows to cause this closure is recorded.

More paste is added, a new groove formed and the test is repeated until two consecutive tests give the same number of blows to cause 13 mm closure of the groove at the base of the cup. About 10 g of the paste is immediately removed from the cup and the moisture content of the paste is determined.

More distilled water is then added to the paste, thoroughly mixed for 10 minutes and the same test procedure is repeated. Thus the moisture content of the paste for each test is increased, the soil flows more easily and the number of blows required for closure of the groove decreases.

At least four tests are required and ideally the four (or more) tests should give a range of number of blows required of between fifty and ten.

A graph of moisture content (as ordinate) versus number of blows (as

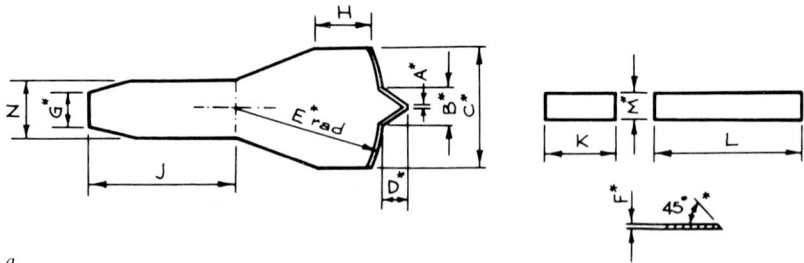

Dimensions						
Letter	A	B	C	D	E	F
mm.	2 ± 0.25	11 ± 0.25	40 ± 0.5	8 ± 0.25	50 ± 0.5	1.5 ± 0.1
Letter	G	H	J	K	L	M	N
mm.	13 ± 0.5	20	50	25	50	10 ± 0.25	20

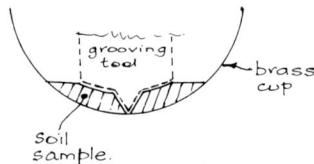

Fig. 2.3 (a) Grooving tool and height gauge (brass or stainless steel)
 *essential dimension
 (b) Section through Casagrande liquid limit cup

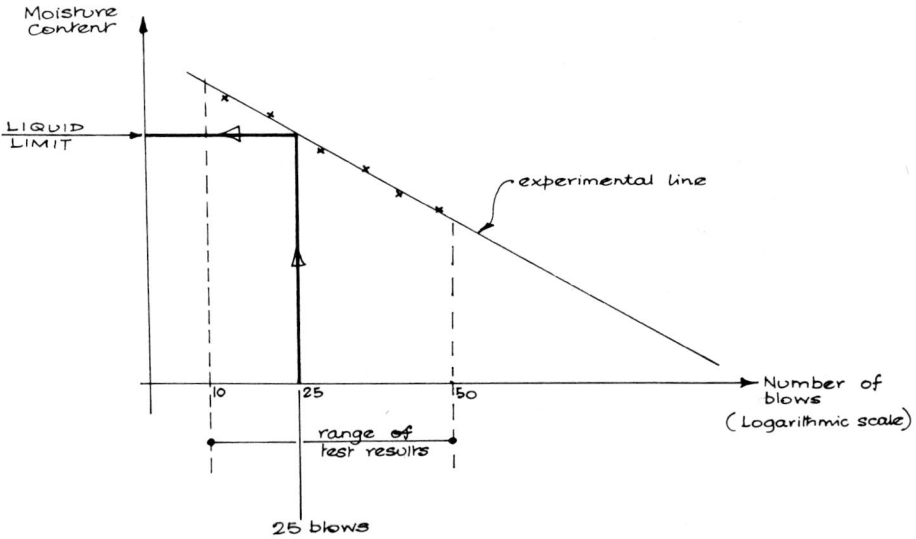

Fig. 2.4 Relationship between moisture content and number of blows

abscissa) to a logarithmic scale is plotted (Fig. 2.4). The liquid limit is taken as the moisture content (expressed to the nearest one per cent) which corresponds to twenty-five blows.

Typical experimental results, together with other reported details, are given in Datasheet No. 1.

Method Using Cone Penetrometer
The penetrometer is shown in Fig. 2.5. Essentially in this test the penetration of a standard cone into a soil sample is measured at a variety of moisture contents and the moisture content corresponding to a penetration of 20 mm is taken as the liquid limit of the fine-grained soil.

As in the previous method, the thick, homogeneous paste is taken from the airtight container and mixed for 10 minutes on a glass plate with a palette knife. The paste is then carefully pushed with a palette knife into a cylindrical metal cup (55 mm diameter by 40 mm deep), care being taken not to entrap air. The surface of the paste is scribed horizontal with a bevelled straight-edge.

The cone is lowered so that it just touches and marks the surface of the soil and the dial gauge reading is noted. The cone is released to penetrate the surface of the soil for 5 seconds, when a second dial gauge reading is taken. The difference between the initial dial gauge reading and the reading after 5 seconds' penetration is taken as the cone penetration.

Some cones have automatic release and locking devices fitted which allow the cone to be lowered and then locked after the 5-second time-interval. If, however, the cone is manually operated, great care needs to be taken to avoid even the slightest vibration of the apparatus when releasing or locking the cone.

MOISTURE CONTENTS

Test	Test Nº	Tin Nº	Tin + Wet Soil w (g)	Tin + Dry Soil W_D (g)	Water $W_w = W - W_D$	Tin W_t (g)	Dry Soil $W_s = W_D - W_t$	Moisture Content % $m = \dfrac{W_w}{W_s} \times 100$
Liquid Limit	1	61	37.11	31.7	5.41	25.05	6.65	81.4
	2	18	40.13	34.0	6.13	26.68	7.32	83.7
	3	17	38.77	32.3	6.47	24.73	7.57	85.5
	4	73	37.69	31.9	5.79	25.39	6.51	88.9
Plastic Limit	1	13	37.2	34.2	3.00	26.1	8.1	32.7
	2	27	36.4	34.0	2.4	26.7	7.3	33.0
	3	19	40.6	37.8	2.8	29.3	8.5	33.1
	4	82	41.13	38.1	3.03	28.8	9.3	32.6
	5	53	36.39	33.8	2.59	25.9	7.9	32.8
	6	41	36.90	34.2	2.70	26.1	8.1	33.1
	7	39	36.05	33.0	3.05	23.7	9.3	32.8
	8	3	41.96	38.7	3.26	28.7	10.0	32.6
Natural m.c.		97	45.55	39.1	6.45	29.0	15.0	43.0

Datasheet N° 1

DATE OF TEST........*3 Nov. 1977*.... JOB.....*Samal Ltd.*.....
BOREHOLE N°.....*12A*.....   SAMPLE.....*7*.........
NATURAL MOISTURE CONTENT.........*43%*.......
UNIT WEIGHT.........*17.7 kN/m³*

LIQUID LIMIT					
Test N°	Tin N°	Number of blows			Average	Moisture Content m %
1	61	48	48	48	48	81·4
2	18	29	29	29	29	83·7
3	17	20	20	20	20	85·5
4	73	10	10	10	10	88·9

PLASTIC LIMIT		
Test N°	Tin N°	Moisture Content m %
1	13	32·7
2	27	33·0
3	19	33·1
4	82	32·6
5	53	32·8
6	41	33·1
7	39	32·8
8	3	32·6
Average m = 32·83		

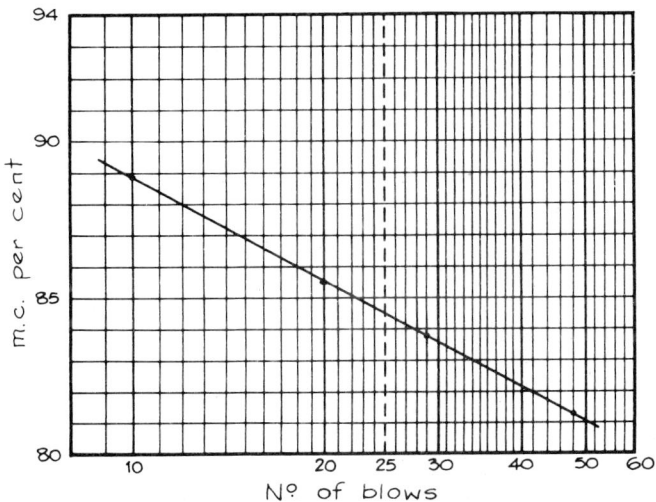

LIQUID LIMIT FLOW CURVE

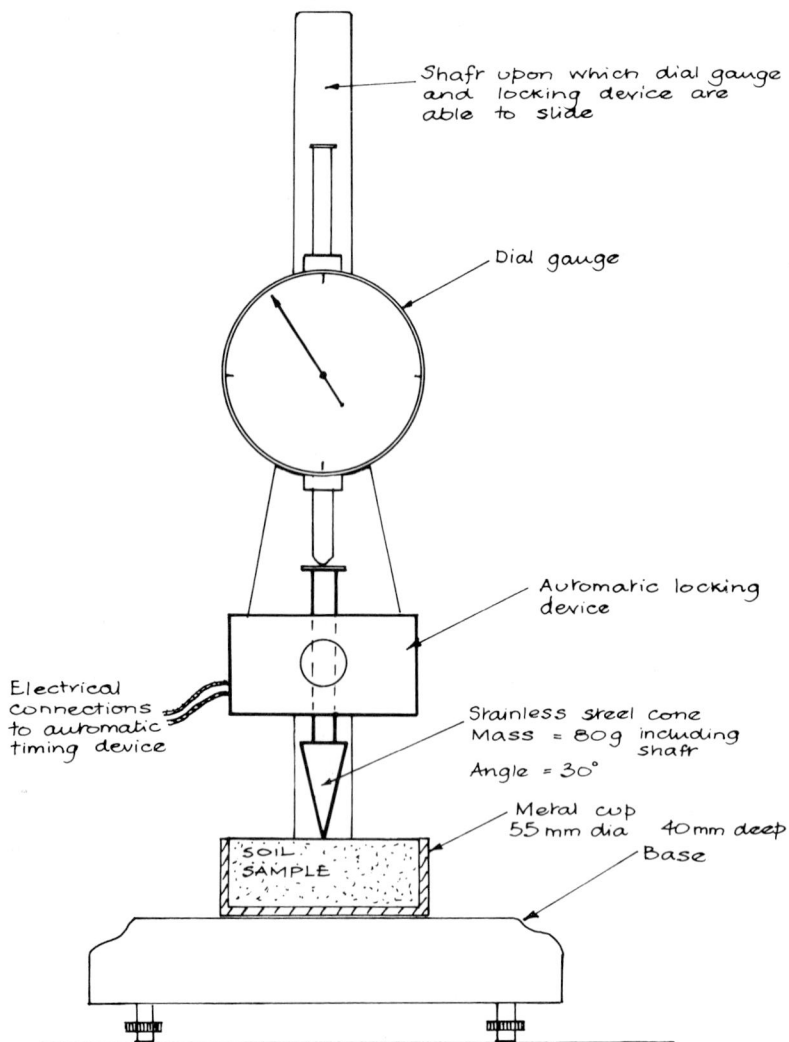

Fig. 2.5 Cone penetrometer for liquid limit test

The cone is then withdrawn and carefully cleaned. More of the soil paste is added to the cup and the test is repeated.

If the difference between the two recorded penetrations is less than 0·5 mm, the average penetration is recorded. If the difference is between 0·5 mm and 1·00 mm, a third test is carried out and, provided that the overall range is then not more than 1·00 mm, the average of the three recorded penetrations is taken. If the range of penetrations is greater than 1 mm, the tests are rejected as inconsistent and the whole process is repeated. As before, a 10 g sample of the paste is taken for moisture content determination which will correspond to the average penetration recorded.

The test is repeated at various moisture contents to give the relationship between cone penetration and moisture content. At least four test results are required and the penetrations should be within the range 15 mm to 25 mm. This means that a 'dummy' run on the soil paste should first be carried out, and if necessary water added and re-mixed so that the initial penetration recorded is at least 15 mm.

A graph of cone penetration versus moisture content is plotted (both to a linear scale) and a straight line should result. The liquid limit is taken as the moisture content at which the standard cone penetrates 20 mm into the soil paste. Datasheet No. 2 displays typical experimental results.

2.2.2 Plastic Limit Test

A representative sample of air-dried soil passing the 425 μm BS test sieve is thoroughly mixed with distilled water on a glass plate until it is sufficiently plastic to be moulded into a ball.

In this test, because of the lack of controlled apparatus and because of the dependence placed upon the person testing, it is essential that a strict uniformity of procedure is adopted.

Initially the ball of soil is rolled and formed between the hands until slight hairline cracks appear at the surface. The ball sample is then split into two samples (about 10 g each). Each of these two samples is then divided into four equal parts and each part (i.e. eight total) is tested as follows.

The soil is rolled between finger and thumb into 6 mm diameter threads and each thread is then rolled between finger-tips and a clean flat glass plate with sufficient pressure to reduce the diameter to 3 mm. The soil is then remoulded into a 6 mm diameter thread and again rolled on the glass plate until of 3 mm diameter. This procedure is repeated until longitudinal and transverse cracks appear at a rolled diameter of 3 mm. Immediately, the moisture content of the cracked thread is determined.

The test is repeated with the other three parts of the 10 g sample and further repeated four times with the other 10 g sample. Satisfactory results are obtained when the means of the two sets of results agree to within 0·5% moisture content.

2.2.3 Summary

The results of liquid-limit and plastic-limit tests carried out on fine-grained soils can be used to aid the process of classifying the soil. By comparing the plasticity index with the liquid limit on Casagrande's plasticity chart, certain tentative conclusions can be drawn about strength, compressibility, plasticity and the type of soil. Important facts about the soil can also be revealed by the liquidity index and activity. References 1, 2 and 3 will prove helpful in making judgements and drawing conclusions from test results and in subsequent determination of other indices.

BS 1377: 1975, *Methods of test for soil for civil engineering purposes*, states

Liquid Limit - Cone Penetrometer Test.

Test	Initial Dial Gauge	Final Dial Gauge	Difference = Penetration (mm)	Mean Penetration (mm)
A1	20·00	4·32	15·68	15·63
A2	19·98	4·41	15·57	
B1	20·00	2·52	17·48	17·25
B2	19·71	2·69	17·02	
C1	24·00	3·29	20·71	20·51
C2	23·90	3·59	20·31	
D1	24·00	1·27	22·73	22·36
D2	23·75	1·62	22·13	
D3	24·04	1·83	22·21	

Test	Tin Nº	Mass of Tin (g)	Tin + wet soil (g)	Tin + dry soil (g)	Mass of water (g)	Mass of Solid (g)	Moisture Content (%)	Mean m (%)
A1	31	110·0	125·7	121·85	3·85	11·85	32·4	32·5
A2	47	111·2	126·9	123·04	3·86	11·84	32·6	
B1	63	113·1	128·3	124·05	4·14	10·95	38·8	39·0
B2	40	111·0	126·2	121·92	4·28	10·92	39·2	
C1	3	112·3	127·3	122·13	5·19	9·83	52·6	52·8
C2	19	110·7	126·4	120·96	5·44	10·26	53·0	
D1	20	110·9	129·8	122·65	7·15	11·75	60·9	61·2
D2	34	110·0	130·3	122·59	7·70	12·59	61·3	
D3	49	112·0	127·1	136·48	5·73	9·37	61·4	

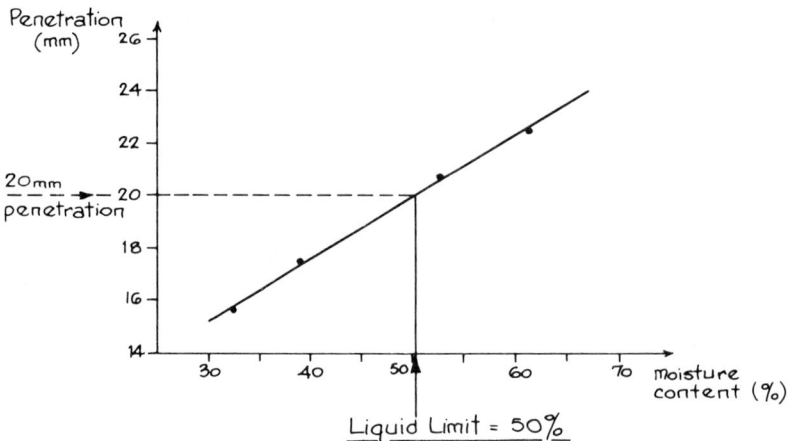

Liquid Limit = 50%

that the cone-penetration test is preferable to the Casagrande apparatus method for the determination of the liquid limit, for the following reasons:

The test is easier to carry out
It gives more reproducible results
The apparatus is easier to maintain in correct adjustment
The test procedure is less dependent upon the judgement of the operator.

Any differences between the results of the two methods are insignificant, being usually very slight.

With the cone-penetration test, care should be taken to ensure

(a) that no air is trapped when forming the test sample in the container;
(b) that vibration or jerking of the apparatus is avoided at all times;
(c) that timing of the fall of the cone, with non-automatic devices, is accurately controlled to 5 seconds;
(d) that the test is always carried out from the dry to the wet state;
(e) that thorough cleaning and drying of the cone and container is carried out at the appropriate stages.

With the Casagrande apparatus, the major difficulty is maintenance of the apparatus so that it completely conforms to a standard procedure in each test. This is particularly important with regard to the rise and fall of the brass cup. Other possible drawbacks are, with some soils, the inability to form a satisfactory groove, the groove closing at two points rather than over a continuous 13-mm length, and the soil 'sliding' on the cup rather than 'flowing' into the groove.

Again, clean apparatus and careful vibration-free use are necessary and care must be taken to ensure that the sample does not dry out between tests.

Notwithstanding the above comments, with an experienced operator using carefully maintained equipment, the Casagrande apparatus can produce excellent results and is still widely used in commercial testing.

In the test to determine the plastic limit, the main requirements are to be consistent in the procedure adopted and to attempt to ensure that any finger-pressures applied are sensibly constant. For example, when approaching the 3 mm size of thread, pressure should neither be relieved nor increased to attain the necessary diameter. Cleanliness is again important. The accuracy of this test so much depends upon the person carrying it out that really there is no substitute for experience and uniformity of procedure.

2.2.4 Linear Shrinkage Test

For some soils, particularly those with low percentage content of clay minerals, it may prove difficult to establish the plastic and liquid limits. With such soils an estimate of the plasticity index can be made by using the following expression:

$$\text{plasticity index} \simeq 2 \cdot 13 \times \text{linear shrinkage}$$

where both the plasticity index (PI) and the linear shrinkage (LS) are expressed as percentages.

The procedure is that about 150 g of soil passing the 425 μm BS test sieve is thoroughly mixed with distilled water to form a homogeneous paste. The moisture content at mixing should approximate to the liquid limit of the soil. The soil paste is placed into a brass mould of semicircular cross-section 140 mm long × 12·5 mm radius. Care is taken not to entrap air and gentle vibration may remove any air during forming. The surface of the soil is scribed level and the exposed wall thickness of the mould is carefully wiped clean.

The soil is then air-dried at 60 C until it has shrunk clear of the mould. Drying at 105 C will then complete shrinkage.

After cooling, the sample length is measured. If a slight curvature has developed during drying, a mean dimension is chosen. The percentage linear shrinkage is given by

$$\text{LS} = 1 - \left(\frac{\text{length after drying}}{\text{initial length}}\right) \times 100\%$$

The initial length is usually 140 mm. The mould is shown in Fig. 2.6 and typical results are given in Datasheet No. 3.

2.3 Determination of the Particle-size Distribution of a Soil

The particle-size distribution of a soil is the major classification test, and a knowledge of the distribution will also prove helpful in making a number of other engineering judgements about a soil and has many applications in foundation engineering. The methods employed are fundamentally scientific but it is important that the engineer or technician performing the tests understands the basic principles underlying sizing analyses.

Soil particles within a soil mass are irregular in shape. In most sizing methods the soil particles are considered as 'equivalent spheres'. This assumption is held to be reasonable for particles greater than about 0·005 mm; below this size, microscopic examination has revealed that soil particles are more platelike, with a length-to-thickness ratio varying from, say, 5 : 1 to 300 : 1.

Soils with coarse grains (e.g. sands and gravels) are subjected to sieve sizing-tests and the standard sieve sizes are based upon the size of spheres

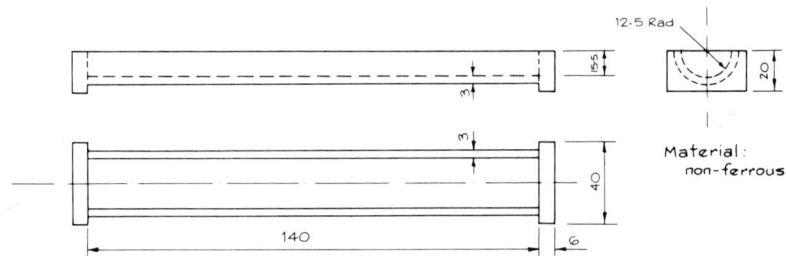

Fig. 2.6 Linear shrinkage apparatus (dimensions in mm)

Datasheet N° 3

Linear Shrinkage Test.

Plasticity Index (%) ≈ 2·13 × (Linear Shrinkage (%)).
Length before drying = 140 mm standard.
Length after drying = 120·3 mm.

$$\text{Linear shrinkage} = \left(1 - \frac{120\cdot3}{140}\right) \times 100$$

$$= 14\cdot07\%$$

Plasticity Index = 2·13 × 14·07 = 29·97 %

that will pass through the sieve mesh. With fine-grain soils the sizing-test is based upon sedimentation tests which are related to spheres falling through a fluid. The methods described below are those most commonly used in civil engineering work.

More sophisticated tests, which may involve microscopical examination, are used to study the effect of the particle shape on the engineering properties of a soil. Microscope work can include the use of electron microscopes and the production of microphotographs.

2.3.1 Sieve Analysis

BS 1377: 1975, *Methods of test for soil for civil engineering purposes*, gives details of two sieve methods, wet sieving and dry sieving.

Paragraph 2.7.2.1 of BS 1377 states that 'the method of dry sieving shall not be used unless it has been shown that for the type of material under test it gives the same results as the method of analysis by wet sieving. In cases of doubt the method shall not be used.'

Both methods allow the particle-size distribution curve of a soil to be determined down to the fine sand size (i.e. 0·06 mm diameter). The procedures are described below and typical results for both methods are given in Datasheet No. 4.

Wet Sieving

Soil is wet sieved to remove the clay- and silt-sized particles ($<0\cdot06$ mm), oven-dried at $105°–110°$ C and then dry sieved to determine the percentage proportion of coarser particles ($>0\cdot06$ mm).

An oven-dried representative sample is weighed to within $0\cdot1\%$ and then placed on the 20 mm BS test sieve. Any particles too large to pass this sieve are wire-brushed clean of any finer material and then sieved on the larger BS sieves, the mass retained by each sieve being noted.

The material passing the 20 mm sieve is riffled to about 2 kg and its mass is noted. It is then spread out in a large tray and covered with water. For each litre of water 2 g of sodium hexametaphosphate (commercial grade) is

Sieve Analysis.   Wet / Dry.

Original mass of sample = 3573 g.

BS Sieve Size	Mass Retained (g)	Sum Mass Retained (g)	% Sum Retained	% Passing	Sieve Size
75 mm	—	—	—	100	75 mm
63 mm	—	—	—	100	63 mm
50 mm	—	—	—	100	50 mm
37.5 mm	60.5	60.5	1.7	98.3	37.5 mm
28 mm	20.9	81.4	2.3	97.7	28 mm
20 mm	60.6	142.0	4.0	96.0	20 mm
Pan ⟶ 3431 g *					

*This mass (3431 g) is now riffled to about 2 kg.
In this example, riffled to 2107 g. Any subsequent
mass retained must be multiplied × $\dfrac{3431}{2107}$ to correct
for the riffling process.

Sieve Size	Mass Retained (g)	Corrected Retained (g)	Sum Retained (g)	% Sum Retained	% Passing	Sieve Size
14 mm	217.4	354.0	496	13.9	86.1	14 mm
10 mm	171.5	279.3	775.3	21.7	78.3	10 mm
6.3 mm	266.2	433.5	1208.8	33.8	66.2	6.3 mm
Pan ⟶ 1451.9 g *	2364.3					

*This mass is again riffled down to about 150 g.
In this example riffled to 141 g.   A second correction
is now applied to all subsequent masses retained.
ie. Each mass is multiplied × $\left(\dfrac{3431}{2107} \times \dfrac{1451.9}{141} \right)$

Sieve Size	Mass Retained (g)	Corrected Retained (g)	Sum Retained (g)	% Sum Retained	% Passing	Sieve Size
5 mm	35.2	590.2	1799	50.4	49.6	5 mm
3.35 mm	31.9	534.9	2333.9	65.3	34.7	3.35 mm
2 mm	28.9	484.6	2818.5	78.9	21.1	2 mm
1.18 mm	10.5	176.1	2994.6	83.8	16.2	1.18 mm
600 μm	8.0	134.1	3128.7	87.6	12.4	600 μm
425 μm	6.3	105.6	3234.3	90.5	9.5	425 μm
300 μm	4.0	67.1	3301.4	92.4	7.6	300 μm
212 μm	8.0	134.1	3435.5	96.2	3.8	212 μm
150 μm	4.1	68.8	3504.3	98.1	1.9	150 μm
63 μm	3.0	50.3	3554.6	99.5	0.5	63 μm
Pan ⟶ 1.1	18.5	3573.1				

The appropriate particle size distribution is plotted overleaf.

Datasheet N° 4

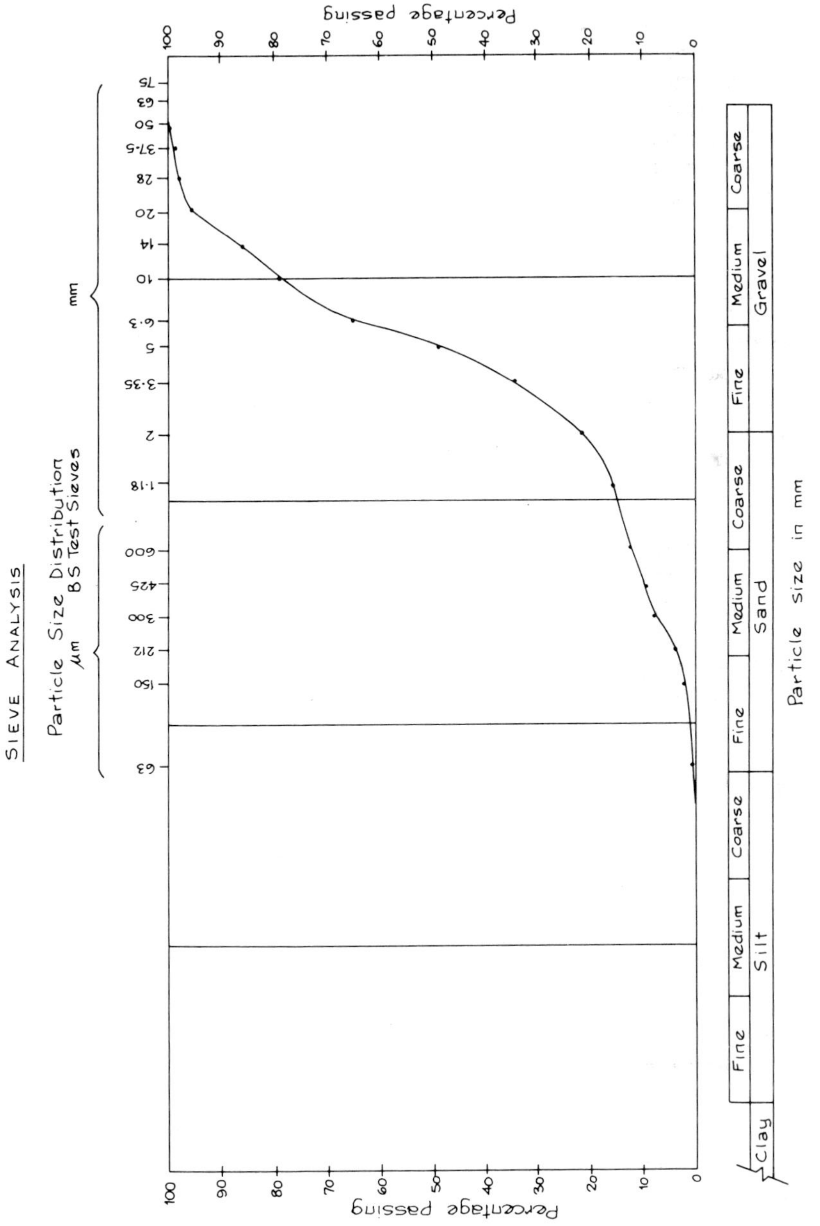

added, and the whole is thoroughly stirred to wet the soil completely. After having been stirred frequently and under water for 1 hour, the soil is washed, little by little, through a 2-mm BS sieve resting on a 63-μm BS sieve. At no stage should either sieve be overloaded (see Table 2.2). After all the material is washed, that retained is oven-dried at 105°–110° C, then it is sieved dry through the appropriate sieves (Table 2.2) down to the 6.3 mm sieve size.

Table 2.2   Maximum loads for sieves of diameter $\leqslant 300$ mm

BS sieve size	Maximum load on sieve
50 mm	4·5 kg
37·5 mm	3·5 kg
28 mm	2·5 kg
20 mm	2·0 kg
14 mm	1·5 kg
10 mm	1·0 kg
6·3 mm	750 g
5 mm	500 g
3·35 mm	300 g
2 mm	200 g
1·18 mm	100 g
600 μm	75 g
425 μm	75 g
300 μm	50 g
212 μm	50 g
150 μm	40 g
63 μm	25 g

Sieving then proceeds down to the 63 μm size noting that, if less than 150 g passes the 6·3 mm BS sieve, then the whole sample is sieved down to 63 μm; but if more than 150 g passes the 6·3 mm BS sieve, this must be riffled down to about 100/150 g before further sieving down to the 63μm size.

At each stage the appropriate mass retained is recorded and expressed as a percentage of the mass of the whole sample.

Dry Sieving
In dry sieving, the soil is sieved down the range of appropriate BS test sieves and the amount retained by each sieve is noted.

Summary of Sieve Analysis
Sieve analysis employed as a sizing analysis can be reasonably accurate so long as its limitations are recognised. The sieves should be well maintained and never overloaded since this leads to clogging of the mesh. The sieving should be accompanied by lateral and vertical movements together with slight jolting. Although sieves are of standard sizes, methods of sieving are far from standard so it is imperative, for reliable results, to standardise the method. Particularly important are:

mode of sieving
period of sieving
sample size.

To standardise the mode and time period of sieving, mechanical shakers are useful and allow the sieving process to be carried out more easily. Sieving should continue for at least 10 minutes and care is needed to ensure that sieving is complete.

2.3.2 Sedimentation Test for Fine-grained Soils

Sedimentation is the process whereby a steady fall of particles occurs through a liquid at rest. The particle sizes are determined from Stokes's law which relates velocity of a spherical particle falling through a liquid to the radius of the particle, the specific gravity of the particle and the viscosity of the liquid.

For a spherical particle,

$$V = \frac{H}{T} = \frac{2\,(\gamma_s - \gamma_L)g \cdot r^2}{9 \qquad \eta}$$

where V = velocity of falling particle (cm/s)
H = distance (cm) through which the particle falls in time T(s)
g = acceleration due to gravity (cm/s^2)
r = radius of the particle (cm)
η = viscosity of the liquid (poise)
γ_s = specific gravity of the particle
γ_L = specific gravity of the liquid.

The velocity of the particle should be limited so that turbulent flow is not created in the suspension and this condition is satisfied if the Reynolds number for the system does not exceed 0·2, at which value Stokes's equation has an error of about 5%. In terms of particle size this means an upper size limit of about 0·2 mm equivalent sphere diameter. Stokes's law can be modified to allow for the non-spherical nature of the particles but, for most practical foundation problems, consideration of equivalent sphere diameters is satisfactory.

Sedimentation testing is best carried out using a pipette (Fig. 2.9); the full test apparatus is shown in Fig. 2.11. The detailed procedure is as follows.

Pretreatment of Soil (Figs 2.7, 2.8)
It is important to ensure that the soil particles are free to fall as individual particles under gravity and that no restraint is placed upon the particles.

Organic matter within a soil tends to be adhesive and aggravates the particles so is removed by hydrogen peroxide.

Calibration of the Sampling Pipette
The type of pipette usually employed is shown in Fig. 2.9. It is lowered into distilled water with tap 1 closed and tap 2 open. Water is sucked into the pipette until it is higher than tap 2, then tap 2 is closed, and the surplus water run off above tap 2 via tap 3.

In a 650 cc Conical beaker Add 50cc. of distilled water

Air-dried soil (12g to 30g) → Riffle Box → (conical beaker) → Boil to reduce volume to 40 cc → Cool

After vigorous frothing reduce volume, by boiling, to 50 cc. ← Gently heat. Avoid frothing over. Stir at frequent intervals ← Cover with glass plate. Allow to stand overnight. ← Add 75cc. Hydrogen Peroxide (20 vol. solution)

In a centrifuge bottle of known mass, make volume up to 200cc. Add Stopper. Centrifuge at 2000 rev/min for 15 mins. Decant clear liquid. → Oven-dry overnight → Cool in a dessicator

Re-weigh contents of bottle hence mass of pretreated soil

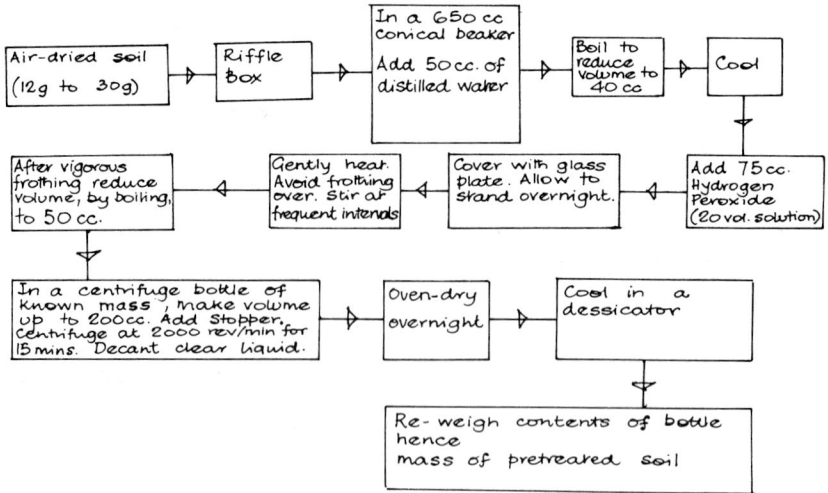

Fig. 2.7 Removal of organic matter by the use of hydrogen peroxide

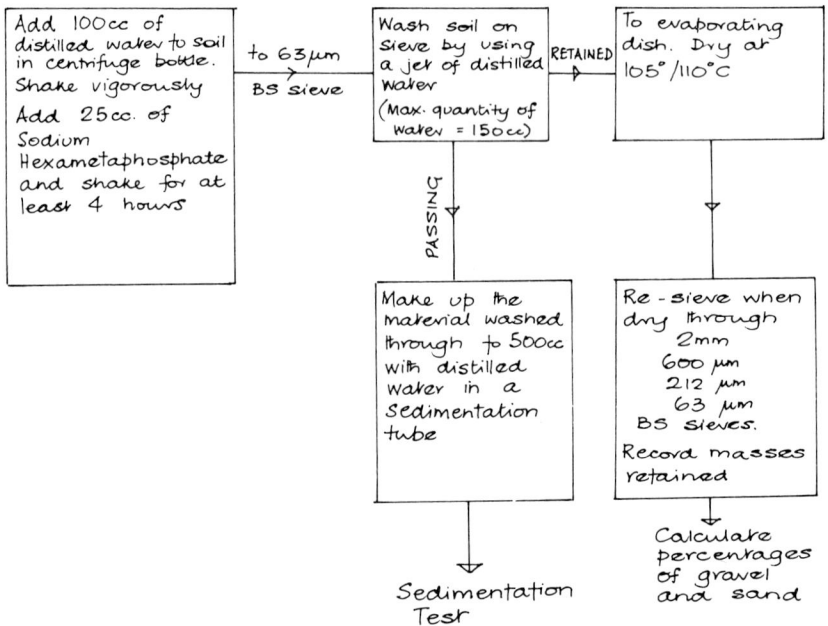

Add 100cc of distilled water to soil in centrifuge bottle. Shake vigorously

Add 25cc. of Sodium Hexametaphosphate and shake for at least 4 hours

→ to 63 μm BS sieve →

Wash soil on sieve by using a jet of distilled water (Max. quantity of water = 150cc)

RETAINED → To evaporating dish. Dry at 105°/110°C

PASSING

Make up the material washed through to 500cc with distilled water in a sedimentation tube

Re-sieve when dry through 2mm 600 μm 212 μm 63 μm BS sieves. Record masses retained

↓ Sedimentation Test

Calculate percentages of gravel and sand

Fig. 2.8 Dispersion of soil

The water held in the pipette and tap 2 is then run into a clean dry weighing bottle. If the masses of the weighing bottle and the weighing bottle + water are known, the volume of the pipette can be determined accurately to the nearest 0.05 cm^3. It is usual to take the mean of several observations.

Sedimentation Process
A suspension of the pretreated soil of known concentration is made up, in which initially all the soil particles are uniformly distributed. The particles of the soil are then allowed to fall freely through the suspension.

From Stokes's law it is possible to determine the time T after which all particles of diameter D will have settled through a distance H. Therefore, when time T from the start of sedimentation has elapsed only those particles with a diameter *less than* D remain at depth H.

For example, suppose that an amount x cm^3 of the suspension is drawn off

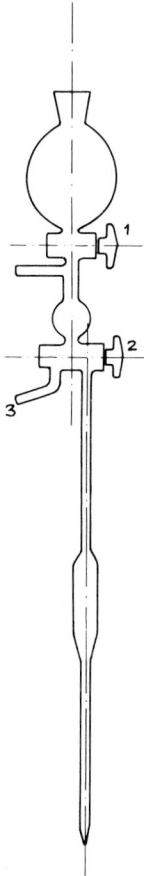

Fig. 2.9 Andreassen pipette

at a depth H in the suspension container after a time T. This amount x cm^3 is then evaporated to dryness and the mass W of the solid matter after drying is recorded. If Y is the weight of soil particles originally used to make up 1000 cm^3 of the suspension, then the percentage of particles having a grain size less than D is

$$\frac{\text{(mass of particles/ml) at depth } H \text{ after time } T}{\text{(mass of particles/ml) in original suspension}} \times 100$$

that is,

$$\frac{W/x}{Y/1000} \times 100 = \% \text{ of particles present of grain size less than } D$$

If the fixed depth H is taken as 10 cm, using Stokes's law the corresponding sampling times and diameters for medium silt, fine silt and clay are as follows:

$T = 4$ min 5 s	$D = 0.02$ mm
$T = 46$ min	$D = 0.006$ mm
$T = 6$ h 54 min	$D = 0.002$ mm

These times assume that the specific gravity of the solid particles is 2·65, that the suspension is made up with distilled water and that the suspension is kept at 20° C throughout the test period (this is because viscosity varies with temperature).

The test procedure is outlined in Fig. 2.10. A completed calculation together with plotted results is given in Datasheet No. 5.

Summary of Pipette Method
Points to note about the application of Stokes's law:

1. Its range of applicability is for particles having equivalent particle diameters of between about 0·2 mm and 0·0002 mm, falling through water. For larger particles, turbulence occurs around the falling particles, while with smaller particles there is a risk of Brownian movements.

2. The equivalent sphere size is obtained from the analysis but the particles may not resemble spheres.

3. The law assumes an infinite extent of liquid around the particle and that the free fall of a particle is not affected by other particles. In soils-laboratory work the suspension tubes are usually of 50 mm diameter and the walls of the tube may affect the finer particles. So long as the concentration of the suspension is maintained at less than 50 g per 1000 ml and the suspension tubes are in excess of 50 mm diameter, any errors are generally negligible.

4. The value used for the specific gravity of the solid particles is an average value and may well be in error because no allowance is made for adsorbed water films around particles. (With some clays adsorbed water can amount to two-thirds of the volume of the clay particle, so that the true specific gravity of the solid matter is less.) For consistent results it is neces-

500 cc. tube. Add 25cc. of sodium hexametaphosphate and make up to 500 cc with distilled water	Add stopper. Stand for 1 hour in constant temperature bath	Remove tubes from bath. Shake thoroughly and invert ten times. Replace in bath and start stop-clock at the same time

Take sample at the specified time, taking 10 secs. to sample. Carefully remove pipette	Lower pipette carefully into suspension tube 15 secs. before specified sampling time. Lower at 1 cm/sec.	

Wash all contents of pipette tube and tap 2 into weighing-bottle of known mass	Oven-dry at 105°/110°C	Determine mass of solid matter in suspension sample

NOTE.

At any intermediate time, use pipette to sample the sodium hexametaphosphate solution from any depth	Determine amount of solid matter in sodium hexametaphosphate sample

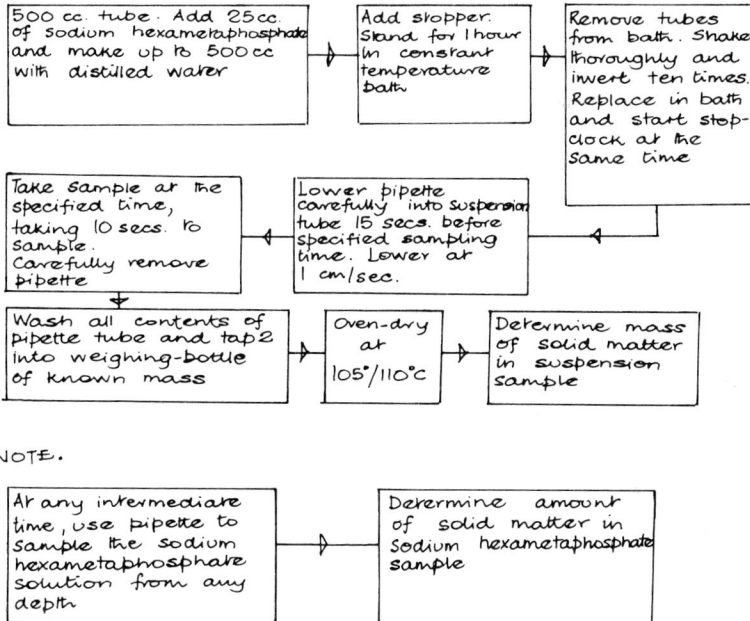

Fig. 2.10 Sedimentation test procedure

sary to keep drying temperatures within the standard range of 105°–110° C, i.e. avoid high-temperature drying.

5. It has been shown with certain soils that putting them into suspension may change their particle size characteristics. For example, stirring can break down some illite and montmorillonite mineral particles into smaller particles.

6. Laboratory temperature variations can cause convection currents within the suspension.

7. Some soils may not disperse easily and other dispersing agents may be required.

In conclusion, however, it is usually sufficient for most purposes to determine the approximate proportions of the silt- and clay-size particles and, in fact, the engineering properties of such soils very often depend upon other factors such as plasticity and mineral type.

Another method employed to determine the silt and clay fractions of a soil is the hydrometer method. BS 1377: 1975 gives full details of this method but classes the pipette method as the primary method for particle size analysis of fine-grained soils. Obvious comparisons between the two methods can be made on the basis of cost, delicacy of equipment, suitability for laboratory or site control tests, simplicity and convenience, but experience over many years tends to favour the pipette method.

Particle Size Analysis - Pipette Method.

Original mass of dry soil   = 45.41 g
Pretreatment with Hydrogen Peroxide
Mass of dry soil after pretreatment   = 43.06 g
                    Pretreatment loss =   2.35 g

Soil dispersed in 20cc of sodium hexametaphosphate.
Suspension made up to 500 cc.
Initial concentration = $\dfrac{43.06}{500}$ = .08612 g/cc.

Pipette calibrated volume = 10.04 cc.
Specific gravity of Solid Particles = 2.65
Test temperature = 20°c throughout.

Sampling times. t_1 = 245 seconds   (dia. = 0.02 mm)
                  t_2 = 46 minutes (dia. = 0.006 mm)
                  t_3 = 6 hours 54 minutes (dia. = 0.002 mm)

(Note. These times correspond only to values of G_s = 2.65
       and a temperature of 20°c.)

Sampling Time		t_1=245 s	t_2=46m	t_3=6h.54m
Depth of sample	(cm)	10.00	10.00	10.00
Bottle number		27	19	3
Mass of Bottle	(g)	27.312	26.481	25.290
Bottle + dry soil	(g)	27.629	26.725	25.503
Dry soil	(g)	0.317	0.244	0.213
Conc. of sample C_s	(g/cc)	$\frac{0.317}{10.04}$=.0316	.0243	.0212

Sample of sodium hexametaphosphate taken by pipette to
determine solid matter per unit volume in the dispersing agent.
   Mass of bottle (g)   = 26.137
   Bottle + dispersing agent (g) = 26.153
   Dispersing agent concentration $D_c = \dfrac{.016}{10.04}$ = .0016 g/cc.

   True concentration of soil solid = $C_s - D_c$
      Sample 1 = .0316 - .0016 = .0300 g/cc.
      Sample 2 = .0243 - .0016 = .0227 g/cc.
      Sample 3 = .0212 - .0016 = .0194 g/cc.

Percentage of particles remaining in suspension at
the sampling times = $\dfrac{(C_s - D_c)}{\text{Initial concentration}}$ × 100%

Particle Size Analysis.

$t_1 = 245\,s$ (dia. = ·02 mm)   Coarse / Medium Silt

% remaining $= \dfrac{0.0300}{0.08612} = 34.84\%$

$t_2 = 46\,min$ (dia. = ·006 mm)   Medium / Fine Silt

% remaining $= \dfrac{0.0227}{0.08612} = 26.36\%$

$t_3 = 6h.\ 54\,min$ (dia. = ·002 mm) Fine Silt / Clay

% remaining $= \dfrac{0.0194}{0.08612} = 22.53\%$

If all the original soil sample passes the 63 µm BS sieve,
then these percentages are %'s of the whole sample.
If not, then they must be related to the Sieve Analysis.
Assuming the first case, then results plotted below show
that this soil sample is 65·16% Coarse Silt
                          8·84% Medium Silt
                          3·83% Fine Silt
                         22·50% Clay
                based on BS Particles sizes.

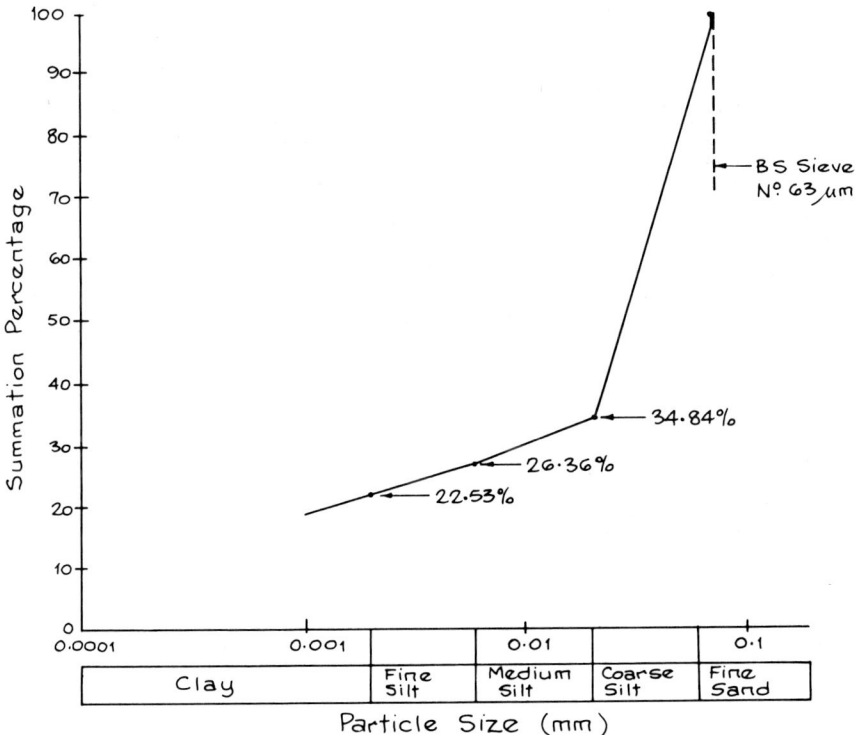

Particle Size (mm)

Fig. 2.11 Particle-size analysis using the pipette method

2.4 Determination of the Specific Gravity of Soil Particles

The specific gravity of solid particles, usually given the notation G_s, is widely used in both laboratory and analytical work. A value of G_s is, of course, required to use Stokes's law in particle-size analysis. Two methods are usually employed: the density bottle method for fine-grained soils; the pycnometer method for coarse-grained material and fine-grained soils.

2.4.1 Density-bottle Method

The apparatus required is shown in Fig. 2.12. The detailed procedure is that a density-bottle complete with stopper is completely dried and weighed to $1/1000$ g (W_1). About 5 to 10 g of soil, passing the 2 mm BS sieve, is riffled from a larger sample, oven-dried at $105°–110°$ C, cooled in a desiccator and placed in the density-bottle. The soil, bottle and stopper are weighed (W_2).

De-aired distilled water is then added to the bottle to cover the soil and the whole is gently subjected to a vacuum in a desiccator. This is carried on for perhaps one hour, avoiding rigorous 'boiling', until no more air is seen to be removed. The bottle is then vibrated and the process repeated until no more air is released. More de-aired distilled water is added to fill the bottle, the stopper fitted and the whole is then left at a constant temperature of $20°$ C for 1 hour. If there is an apparent decrease in water volume in the bottle then more is added and the whole allowed to stand. When there is no apparent decrease in water volume after standing, the bottle stopper, soil and water are weighed (W_3) after the exterior of the bottle has been dried.

The bottle is then thoroughly cleansed, filled with de-aired distilled water and allowed to stand for one hour. If there is apparent decrease in water volume, the bottle is 'topped-up' and allowed to stand for a further period. When there is no apparent decrease in water volume after standing, the bottle, stopper and water contents are weighed (W_4).

The specific gravity of the solid soil particles is found from the following expression:

$$G_s = \frac{W_2 - W_1}{(W_4 - W_1) - (W_3 - W_2)} \tag{2.1}$$

At least two determinations of G_s are required.

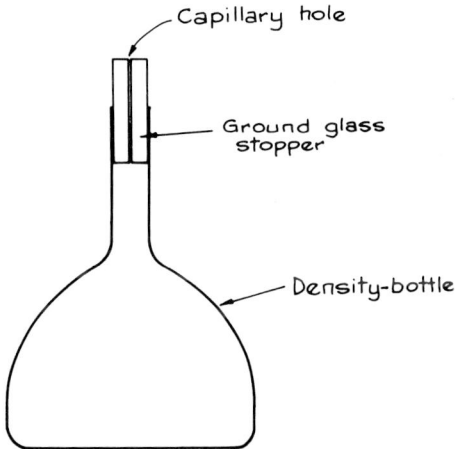

Fig. 2.12 Typical density-bottle

2.4.2 Pycnometer or Gas-jar Method

The choice between using a pycnometer or gas-jar is purely academic. Any container is satisfactory so long as it is capable of being effectively sealed so that it has a constant and consistent volume. The container must also be capable of being mechanically shaken and, obviously, be of sufficient capacity.

A sample of dried soil which has been stored in an airtight container is added to a clean dry pycnometer (complete with sealer) of known mass (W_1) and the soil plus pycnometer and sealer are weighed (W_2). The sample size is usually about 200 g for fine-grained soils and 400 g for coarse-grained soils.

The pycnometer is then half filled with water and fitted with the seal. With coarse- and medium-grained soils the contents are allowed to stand for 4 hours. After this time (or, in the case of fine-grained soils, immediately after the water is added) the pycnometer is shaken by hand to form a suspension and then mechanically shaken for 20 to 30 minutes. The sealer is removed and any soil sticking to it is carefully washed into the pycnometer; further water is added, first to just below the top of pycnometer and finally to top it up, extreme care being taken not to form a froth or to entrap air. The pycnometer is then sealed and the whole is weighed (W_3) after having been dried externally.

The pycnometer is thoroughly cleansed, dried and filled completely with water and reweighed (W_4).

The specific gravity of the solid particles is then found by using eq. (2.1) and the average value of at least two tests is reported.

Typical results are given in Datasheet No. 6.

2.4.3 Summary

The major points to note when carrying out either test are as follows:

1. Accuracy of weighing is vital and careful techniques and procedures are required.
2. De-aired distilled water must be available and used at all stages in such a way that the risks of entrapping air are minimised.
3. Cleanliness of all equipment is of paramount importance, as is the drying of both the equipment and the soil. Standard drying temperatures should be used wherever possible ($105°$–$110°$ C). However, in some cases, for example with organic soils or where there is a risk of hydration, a lower temperature may have to be used.
4. Testing should be carried out under constant-temperature conditions.
5. In some cases an alternative liquid, such as kerosene, may have to be used, for example with a peat soil where G_s is of the order of 1·5 or so. It is usually necessary to determine the specific gravity of the liquid prior to performing the test. Points to note here are that the expression to determine G_s will alter and that the various drying processes must be carried out at room temperature ($20°$ C) to avoid the risk of ignition of the soil and liquid.

Specific Gravity of Soil Particles

a. Density Bottle.

Mass of density bottle   =   25.631 g   (W_1)
including stopper

Mass of density bottle   =   37.311 g   (W_2)
+ stopper + oven dried
soil

Mass of density bottle +   = 62.590 g   (W_3)
stopper + oven dried soil +
distilled water

Mass of density bottle +   = 55.272 g   (W_4)
stopper + distilled water

$$G_s = \frac{W_2 - W_1}{(W_4 - W_1) - (W_3 - W_2)} = \frac{37.311 - 25.631}{(55.272 - 25.631) - (62.590 - 37.311)}$$

$$= \frac{11.680}{29.641 - 25.279} = \frac{11.680}{4.362} = \underline{2.678}$$

* Repeat test at least twice.

b. Pycnometer or Gas Jar.

Mass of pycnometer + cover = 410.03 g   (W_1)

Mass of pycnometer + cover + = 714.37 g   (W_2)
air dried soil

Mass of pycnometer + cover + = 1031.90 g   (W_3)
air dried soil + water

Mass of pycnometer + cover = 838.63 g   (W_4)
+ water

$$G_s = \frac{W_2 - W_1}{(W_4 - W_1) - (W_3 - W_2)} = \frac{714.37 - 410.03}{(838.63 - 410.03) - (1031.90 - 714.37)}$$

$$= \frac{304.34}{111.07} = \underline{2.74}$$

* Repeat test at least twice.

6. With some forms of density-bottle it may be necessary to dry the bottle by rinsing in acetone and then warm-air drying with a blower.
7. Any vacuum work must be carried out with care and precautions taken to minimise the risk of an implosion.

2.5 Soil Chemical Tests

As more and more development work continues on sites previously thought unsuitable for building then two soil tests have become increasingly important. These are

(i) the determination of the sulphate content of a soil
(ii) the determination of the pH value of a soil.

Both of the tests provide significant information about the probable aggressive quality of soil and, depending upon the results obtained, it may be necessary to specify special materials for construction, or even to alter potential modes of construction.

2.5.1 Determination of Sulphate Content of a Soil

Depending upon the soil type, a quantity of soil is oven dried at 80° C and cooled. It is then weighed to 0·1% of its original mass (m_1) and sieved on a 2 mm BS sieve. After removing stones, the dried soil is crushed to pass through the sieve.

The initial mass of soil used is as follows:

Fine grained $\simeq$ 150 g
Medium grained $\simeq$ 600 g
Coarse grained $\simeq$ 3500 g

The stones are removed since they are assumed not to contain sulphates and the mass of crushed soil passing the 2 mm sieve is recorded (m_2), again to 0·1% of the original mass.

The sample is then riffled down to about 100 g and pulverised so that it passes the 425 μm sieve. Further riffling reduces the sample size to about 10 g.

Important points for attention during this sieving and riffling process are:

(a) take care to prevent loss of fine material
(b) at all stages, thoroughly mix all the soil and prevent segregation during riffling.

The 10 g sample is then dried at a temperature between 75° and 80° C. When weighings taken at 4 hour intervals do not vary by more than 0·1% of the original mass then drying is complete. This drying is usually performed in a weighing bottle.

After cooling, and weighing soil and bottle to 0·001 g, a sample of about 2 g is transferred to a 500 ml beaker and the weighing bottle and contents are re-weighed and mass (m_3) is the difference in mass before and after removal of the soil to the beaker.

Hydrochloric acid (200 ml of 10%) is added carefully to the beaker taking care to retain all the material. The beaker is covered with glass and the contents boiled for 4 or 5 minutes.

The contents are allowed to boil gently, the underside of the glass cover is carefully washed into the beaker and 3 ml of bromine water added whilst the gentle boiling continues.

Ammonia solution is then added carefully from a burette whilst the solution is gently stirred and continues to boil. This is continued until precipitation occurs and ammonia vapour is detectable.

The suspension is then filtered through a Whatman No. 54 filter paper (110 mm) into a 500 ml conical beaker.

When filtration is completed, the filter paper and its contents are transferred carefully back to the original 500 ml beaker. A further 20 ml of 10% hydrochloric acid is added and the contents stirred carefully.

The filter paper is removed and washed with water until all the yellow colouring is removed. These washings are collected and, in the absence of any further 'yellowing', the filter paper is scrapped.

Careful boiling of the contents is, once again, started and further ammonia solution added. Further filtering, through a Whatman 541 filter paper, into the conical beaker with the first precipitation is then carried out.

Using litmus paper and by adding hydrochloric acid, the extract is made slightly acidic. This is then boiled and 25 ml of 5% barium chloride solution is added very slowly whilst stirring. This is best done by a pipette drop by drop. The solution is then covered for about one hour whilst remaining hot.

After the solids have settled, a few more drops of barium chloride are added to make sure that full precipitation of barium sulphate has occurred, if not then further barium chloride is added until full precipitation is achieved.

The precipitate is then transferred very carefully to a filter funnel fitted with a Whatman 44 filter paper and filtered. Washing with warm distilled water is then effected.

A porcelain crucible of known mass and already ignited receives the filter paper and the precipitate. The crucible is then heated to about 800° C (red heat) care being taken to do this gradually so that the filter paper is charred rather than inflamed. This red heat source is maintained for about one-quarter of an hour. After cooling, a few drops of hydrochloric acid and then sulphuric acid (both concentrated) are added and then further ignition (for about one-quarter of an hour) and cooling in a desiccator are undertaken.

The crucible and contents are then re-weighed and the increase in mass is recorded to the nearest 0·001 g (m_4).

The percentage of sulphate present in the soil sample is then calculated using:

$$\% \text{ of } SO_3 = \frac{34\cdot3 \times m_2 \times m_4}{m_1 \times m_3} (\%)$$

where m_1 = mass of sample prior to sieving
$\quad\quad m_2$ = mass passing 2 mm sieve
$\quad\quad m_3$ = mass of soil used
$\quad\quad m_4$ = mass of ignited precipitate.

The SO_3 content of the soil, or sulphate content, is normally reported to the nearest $0\cdot01\%$.

Reference to standard concrete, and other construction material, data will then specify the precautions to be taken to safeguard against possible sulphate attack. For concrete work this may involve using, for example, sulphate-resisting cement and varying the mix proportions from those allowed in non-sulphate ground conditions.

A further test procedure is given as Test 10 in BS 1377: 1975 to determine the sulphate content of groundwater.

2.5.2 Determination of the pH Value of a Soil (Suspension)

After air drying, the sample is crushed to pass through the 3·35 mm BS sieve and, by dividing, the material passing the sieve is reduced to about 30 g. The sample of soil is accurately weighed.

This sample, together with 75 ml of distilled water, is placed in a 100 ml beaker and well stirred.

Normally the suspension is left to stand covered overnight and then re-stirred the following morning immediately prior to testing.

The portable pH meter which has previously been calibrated is then used. The electrodes are washed initially with distilled water and then these electrodes are immersed into the prepared suspension. Three or so readings of the pH value of the soil suspension are taken with stirring between each observation. These readings ought to agree to within $\pm0\cdot05$ pH units.

The electrodes of the portable pH meter are then re-washed with distilled water and its calibration is re-checked.

This method (electrometric method) is usually performed relatively easily and gives good results so long as the meter is maintained to the manufacturer's instructions.

Results are direct reading and are recorded to the nearest 0·1 pH value.

Acidic soils are those having pH values less than 7·0, possibly as low as 4·0, whilst alkaline soils have pH values in excess of 7·0. The pH value of 7·0 defines the neutral condition.

With both of the above tests it is important to realise that both pH value and sulphate content, especially, are subject to variation with the climatic seasons and therefore any results are relative to the time of sampling only.

CHAPTER 3

Seepage and Permeability Tests

Determination of the coefficient of permeability using

    Falling-head permeameter
    Constant-head permeameter
    Field pumping tests
    Rowe consolidation cell.

3.1 Introduction

Initially it is important to realise the difference between porosity and permeability.

Porosity
Porosity expresses the ratio of the void (air and/or water) space of a soil to the total volume (including solid) of the soil (Fig. 3.1).

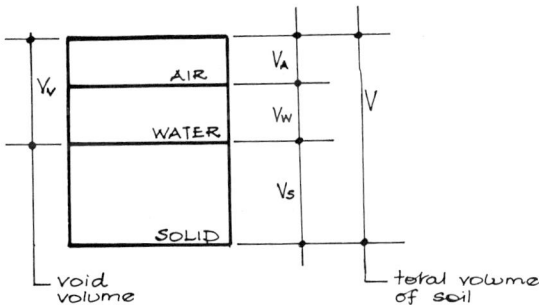

Fig. 3.1

$$\text{Void ratio} = e = \frac{V_v}{V_s}$$

$$\text{Porosity} = n = \frac{V_v}{V}$$

$$V = V_v + V_s$$

so

$$n = \frac{V_v}{V_v + V_s} = \frac{V_v/V_s}{(V_v/V_s) + 1}$$

but

$$e = \frac{V_v}{V_s}$$

therefore

$$n = \frac{e}{1 + e}$$

Note: 1. With a saturated soil $V_a = 0$
2. With a dry soil $V_w = 0$
3. The void ratio e can be >1 or <1.
4. The porosity is always less than 1. It must also be realised that the porosity is a dimensionless quantity.

Permeability
Permeability is a measure of the capacity of a soil to permit the passage of water and has the dimensions of velocity (e.g. m/year, mm/second).

A soil with a high porosity is not necessarily highly permeable. For example, a clay with a moisture content of, say, 20% and having $G_s = 2.65$ has a void ratio e, assuming full saturation, of

$$e = mG_s = 0.2 \times 2.65 = 0.53$$

and porosity

$$n = \frac{e}{1 + e} = \frac{0.53}{1 + 0.53} = 0.346$$

while a sand with, say, $G_s = 2.60$ and $m = 10\%$, assuming full saturation, has

$$e = mG_s = 0.1 \times 2.60 = 0.26$$

and porosity

$$n = \frac{e}{1 + e} = \frac{0.26}{1 + 0.26} = 0.206$$

Thus the clay has a higher porosity than the sand (which is usually the case), while the clay is very much less permeable than the sand. If the sand is clean one may expect a permeability of, say, 10^{-3} m/s; if the clay is unfissured, its permeability will be of the order of 10^{-8} m/s.

In 1856 Darcy defined a coefficient of permeability, from a study of water passing through a saturated porous medium (see Fig. 3.2). Darcy's law states that the apparent velocity of flow is directly proportional to the hydraulic gradient:

$$V \propto i$$

$$V \propto \frac{H}{L}$$

$$V = K\frac{H}{L}$$

$$Q = AV = AK\frac{H}{L}$$

where Q = quantity of water flowing in unit time

$$i = \text{hydraulic gradient} = \frac{\text{head loss}}{\text{drainage path length}}$$

K = coefficient of permeability.

It must be noted that Darcy's law applies to *saturated* porous media (i.e. no air is present within the voids). The presence of air can seriously affect permeability tests and if it is present in significant quantities, Darcy's law is no longer valid.

The permeability of a soil is affected by various physical circumstances, listed overleaf.

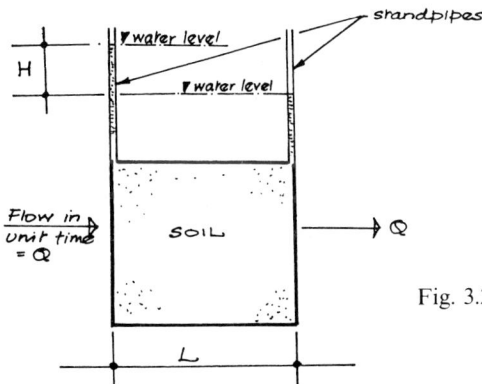

Fig. 3.2 H = head loss over length L
L = length of drainage path
A = area of cross-section of flow
V = apparent velocity of flow

1. Shape and the Arrangement of Particles

The particles within a soil mass can be arranged in a variety of ways. Figure 3.3 shows how the tortuosity of the path of a particle of water can be varied according to the shape and the arrangement of the particles. It must be noted that, while in the analysis the water is assumed to flow one-dimensionally along a straight line, in fact the water twists and turns along a very non-linear path.

2. Void Ratio and Unit Weight

The arrangement of the particles thus affects the void ratio e of the soil and since the porosity n of the soil and the unit weight γ are dependent upon the void ratio, it is clear that alteration of either the void ratio or the unit weight increases or decreases the permeability according to the change in arrangement of the particles. A permeability value is only applicable to the state of the soil 'as tested', therefore it is imperative that the condition of the sample is clearly defined in any permeability test report if the results are to have meaning. Of equal importance is the fact that the test sample should be in the same condition of moisture content and unit weight as the mass of the soil at the field scale.

3. Properties of the Permeant Fluid

The viscosity of the water passing through the soil is largely governed by temperature. Under normal conditions there is little temperature variation of the soil water on site below about 2 m below ground level. The temperature is usually in the range 10°–20° C. However, the temperature in the laboratory where permeability tests are being carried out is of the order of 20° C. To ensure accurate observations it is thus essential that

> (a) laboratory permeability tests are carried out in a constant-temperature room;
> (b) a correction is applied to the calculated value of the coefficient of permeability to account for the effect of temperature on the viscosity of the permeant fluid.

The other important consideration with regard to fluid is that it is of little relevance to pass de-aired and/or distilled water through a sample in the

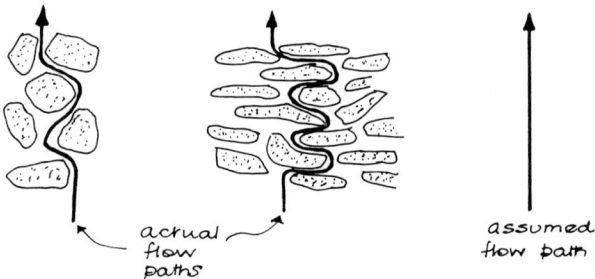

Fig. 3.3

laboratory if it differs from the water found on site. So a refined technique is to use, in the laboratory, water that has been collected from the particular site under consideration. This water is usually filtered before use, and then introduced into the apparatus and associated pipework.

4. Degree of Saturation
If the soil is about 85% saturated, air occurs as bubbles within the pore space of the soil and these can block the seepage channels and thus reduce the permeability of the soil. Under these conditions Darcy's law is only approximately valid. It is important to note that a change of 1% in the degree of saturation above the 85% level can change the permeability by about 3%.

Below a degree of saturation of 85% Darcy's law is invalid and more sophisticated analysis and testing techniques are required. At this level of saturation of the voids, continuous air (rather than bubbles) causes the invalidity of Darcy's law.

5. Type of Flow
Usually the application of Darcy's law requires that the flow is laminar and of such a velocity that the velocity head is negligible when applying Bernoulli's equation to seepage problems.

6. Mineral Characteristics of Fine-grained Soils
Different types of particles attract and hold different thicknesses of adsorbed water around each particle, and consequently reduce the effective pore size.

7. Stratification or Layering
Due to the manner in which soils are deposited, they are usually stratified or layered. Consequently a soil can have markedly different permeability values in horizontal and vertical directions. This effect is known as anisotropy. Generally the permeability value in a horizontal direction (along the layers) is much greater than the permeability value in a vertical direction (across the layers). It has been shown that the difference in permeability values increases with an increase of the effective pressure on the soil.

Other factors that affect permeability include the presence of organic matter, sand lenses, silt intrusions, fissures, etc. This further emphasises the need to test samples that are fully representative of the mass. To attain this end, tests may have to be performed on large samples in an effort to incorporate the many variable features of natural soil deposits.

3.2 Permeameter Tests

3.2.1 The Falling-head Permeameter
This method is usually employed to determine a coefficient of permeability for fine-grained soils. The apparatus is as shown in Fig. 3.4.

The soil sample is usually undisturbed and very often the U4 sampling tube can be used as the container during the test. A coarse filter screen is placed at the upper and lower ends of the sample tube. The base of the sample tube is connected to the water reservoir; to the top of the sample tube is connected a glass standpipe of known cross-sectional area. This pipe is filled with water: as the water seeps down through the soil sample, observations are taken of time versus height of water in the standpipe above base reservoir level. A series of tests is performed, using different sizes of standpipe, and the average value of the coefficient of permeability is taken.

Note must be taken of the unit weight of the sample, and of its moisture content.

Applying Darcy's law,

$$V = -dh/dt \qquad Q = Aki$$

then

$$-\frac{a \cdot dh}{dt} = AK\frac{h}{L}$$

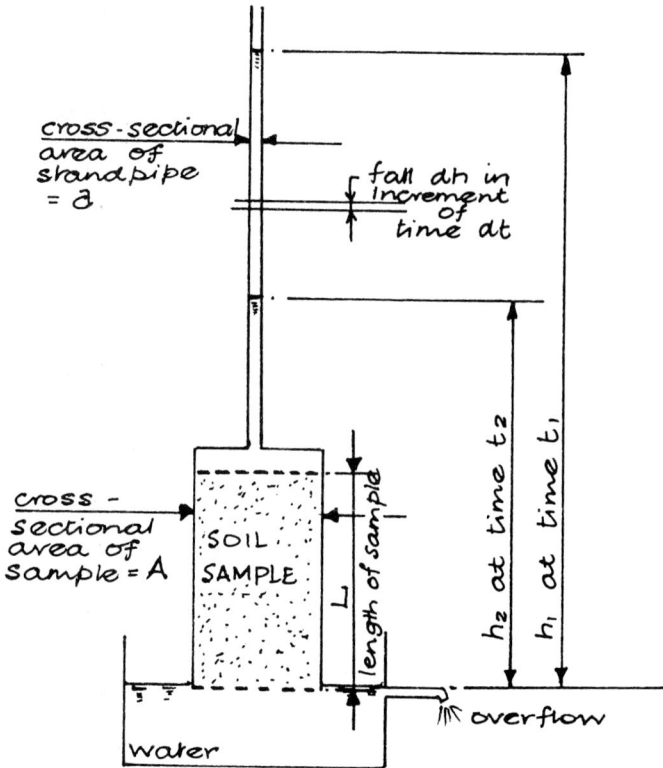

Fig. 3.4 Falling-head permeameter

so, integrating between limits of h_1, t_1 and h_2, t_2,

$$-a \int_{h_1}^{h_2} \frac{dh}{h} = \frac{AK}{L} \int_{t_1}^{t_2} dt$$

$$K = \frac{aL}{A(t_2 - t_1)} \left(-\log_e h_2 + \log_e h_1 \right)$$

$$K = \frac{aL \log_e(h_1/h_2)}{A(t_2 - t_1)}$$

Note. To convert $\log_e$ to $\log_{10}$, multiply by 2·3.

Full results given in Datasheet No. 7 (page 52).

3.2.2 The Constant-head Permeameter

For this test the arrangement of the apparatus is usually as shown in Fig. 3.5. The test is generally used to determine the coefficient of permeability values of coarse-grained soils.

The soil sample is confined within a Perspex cylinder. A wire mesh filter is fitted at the upper and lower ends of the sample. Two connections to the Perspex cylinder allow the loss of head across a fixed length of the sample to be observed. A constant-head supply is connected to the top of the sample and seepage takes place down through the soil; the volume of water passing through the sample is recorded. Usually several tests are performed at varying heads and a mean value of permeability is taken.

It is imperative that the sample unit weight is at the required value and that air is reduced to an absolute minimum. Before starting the test a vacuum can be applied to the sample in an attempt to perform the permeability test at a degree of saturation approaching 100%.

Before taking observations it is necessary to ensure that 'steady state' conditions of seepage exist. This is usually confirmed by running the test overnight. Once the flow rate and the heads from the two tapping points are constant, observations may be recorded. In addition, the cross-sectional area A of the sample and the length L between the tapping points need to be recorded.

The coefficient of permeability is then calculated by simple application of Darcy's law:

$$V = Ki \qquad \frac{Q}{t} = AV$$

therefore

$$\frac{Q}{t} = KAi$$

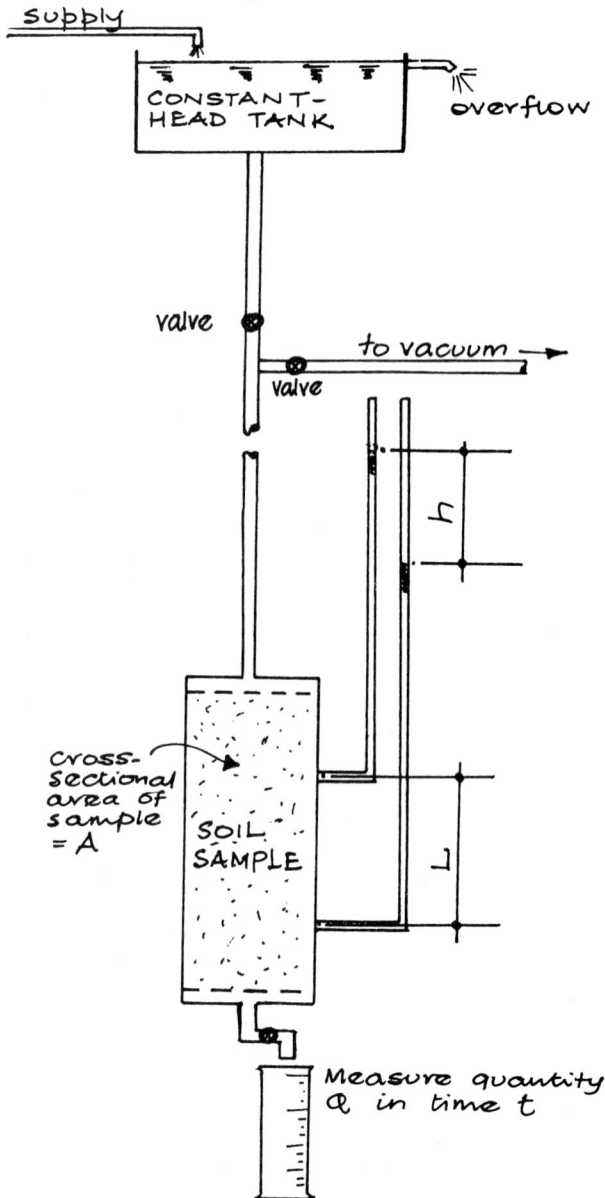

Fig. 3.5 Constant-head permeameter

Now

$$i = h/L$$

therefore

$$K = \frac{QL}{Aht}$$

where Q = total quantity collected in time t
L = distance between tapping points
A = cross-sectional area of the sample
h = difference in level of two standpipes tapped into container.

Typical results are given in Datasheet No. 8.

3.2.3 Summary

At best, both the constant-head and the falling-head permeameter tests as described give only a guide to the coefficient of permeability of the soil under test. The reasons for this are as follows:

1. It is difficult to ensure that there has been no sample disturbance. This is particularly so with coarse-grained soils where, because it is almost impossible to take undisturbed samples, the sample is compacted into the permeameter. Not only must the density of the soil on site be reproduced in the sample, but the arrangement and distribution of particles within the soil should similarly be reproduced. This can hardly ever be achieved in practice.

2. The behaviour of the mass of soil at the field scale hardly ever corresponds to the behaviour of a small sample. Within a large mass of soil there are permeability variations, possibly in every direction, over quite small plan areas. Consequently it is necessary to test representative samples in order to establish valid permeability characteristics. These samples should also incorporate all the features of the soil mass at the field scale.

3. The use of water other than site water may affect the nature of the adsorbed water surrounding fine-grained soil particles and may even lead to chemical action during a permeability test. This may well lead to the production of air (or gas) which will influence measured permeability values.

4. Where the direction of seepage is predominantly horizontal or at an inclined angle, observation of the permeability characteristics in a vertical direction is of little practical value.

5. During the tests the levels of total pressure, pore-water pressure and effective pressure are not usually equivalent to field values and consequently void ratio values do not correspond. It is generally accepted that the coefficient of permeability varies with void ratio. Therefore to obtain reliable data the void ratio under test should correspond to that at the field scale. If the field problem involves a variety of void ratios, corresponding permeability values should be determined and reported.

The results of laboratory permeability tests are frequently augmented or replaced by those of field pumping tests, described below.

1. Falling Head Permeability Test.

Test Nº	Sample Length cm L	Elapsed time h_1 to h_2 $(t_2 - t_1)$ sec	Initial Standpipe Ht h_1 cm	Final Standpipe Ht h_2 cm	$\dfrac{h_1}{h_2}$	$\log_e \left(\dfrac{h_1}{h_2}\right)$	$\dfrac{L}{t_2 - t_1}$
1	24·00	215	150	50	3	1·0986	0·1116
2	18·00	147	150	50	3	1·0986	0·1225
3	19·75	163	150	50	3	1·0986	0·1212
4	22·13	191	150	50	3	1·0986	0·1159

A = Area of sample = 81·07 cm²
a = Area of standpipe = 7·1 cm²

$$k = \frac{a}{A} \cdot L \cdot \frac{\log_e (^{h_1}/_{h_2})}{t_2 - t_1} = \frac{a}{A} \cdot \log_e \left(^{h_1}/_{h_2}\right) \left(\frac{L}{t_2 - t_1}\right)$$

since $\dfrac{a}{A} \log_e \left(^{h_1}/_{h_2}\right) = \dfrac{7·1}{81·07} (1·096)$, then $k = 0.0962 \left(\dfrac{L}{t_2 - t_1}\right)$

Test 1 $k = 1·07 \times 10^{-2}$ cm/s
     2    $k = 1·18 \times 10^{-2}$ cm/s
     3    $k = 1·17 \times 10^{-2}$ cm/s
     4    $k = 1·11 \times 10^{-2}$ cm/s

* Note must also be made of the bulk unit weight of the soil tested, moisture content of all samples, and the temperature at the time of testing.

2. Constant Head Permeability Test.

Test Nº	Quantity flowing through sample Q cc	Time for flow t sec	Difference in standpipe levels h cm	Difference between tapping points L cm	Cross sectional area of sample A cm²	$k = \dfrac{QL}{Aht}$ cm/s
1	500	121	7·31	10	81·07	$k = 6·9 \times 10^{-2}$
2	500	134	6·97	10	81·07	$k = 6·6 \times 10^{-2}$
3	500	129	7·13	10	81·07	$k = 6·7 \times 10^{-2}$
4	500	139	6·88	10	81·07	$k = 6·45 \times 10^{-2}$

3.3 Field Pumping Tests

There is a variety of tests, including those described here; choice depends mainly upon the type and arrangement of the various strata encountered during the site investigation.

3.3.1 Unconfined Flow – Well-pumping Test

The general arrangement is as shown in Fig. 3.6. The following points should be noted:

(a) The hydraulic gradient is assumed to be the slope of the water table or free surface (i.e. $i = dh/dr$) at any radius from the pumping well. It is also assumed to be constant with depth at this radius.

(b) The analysis assumes that the ground level, initial water table and impervious strata are sensibly horizontal.

(c) The flow everywhere is horizontal. This means that the piezometric level at any point coincides with the water table or free surface.

Obviously these assumptions may not be valid.

Applying Darcy's law,

$$V = Ki \qquad Q = AV$$

therefore

$$Q = AKi$$

where A = cross-sectional area of flow = circumference of flow $(2\pi r)$ × depth of flow (h), i = hydraulic gradient = dh/dr, Q = quantity of water discharged in unit time from the pumping well. Therefore

$$Q = 2\pi r h K \frac{dh}{dr}$$

$$\frac{dr}{r} = \frac{2\pi K}{Q} h \, dh$$

Integrating between the limits of $r = r_A$ and r_B and $h = h_A$ and h_B,

$$\log_e \frac{r_B}{r_A} = \frac{2\pi K}{Q} \left(\frac{h_B^2}{2} - \frac{h_A^2}{2} \right)$$

$$K = \frac{Q}{\pi} \left[\frac{\log_e r_B/r_A}{(h_B - h_A)(h_B + h_A)} \right]$$

noting that h is measured to the free surface from the impervious strata (i.e. $h = H$ − depth to WT below GL).

The procedure is that a pumping well is sunk to the base of the deposit under test, fully penetrating the deposit to the impervious strata below. A series of observation wells is then installed around the pumping well, on lines radiating from it, with at least two observation wells per radiating line. As pumping proceeds, the free surface is drawn down towards the base of the pumping well (Fig. 3.6) and, when steady state conditions are reached, observations are taken down to the water level.

The coefficient of permeability is then calculated for each pair of observations on a radiating line, and average values are calculated.

3.3.2 Confined Flow – Well-pumping Test

An alternative test applies to the case of a confined aquifer, i.e. an aquifer confined at its upper and levels levels by relatively impervious strata (Fig. 3.7).

Everywhere the flow is horizontal, radially inwards to the pumping well and of depth H. As before,

$$Q = AKi = 2\pi r H K \frac{dh}{dr}$$

Separating variables,

$$\int_{r_A}^{r_B} \frac{dr}{r} = \int_{h_A}^{h_B} \frac{2\pi K H}{Q} \, dh$$

$$\log_e r_B/r_A = \frac{2\pi K H}{Q}(h_B - h_A)$$

$$K = \frac{Q}{2\pi H} \times \frac{\log_e r_B/r_A}{h_B - h_A}$$

Note that this differs from the expression for unconfined flow.

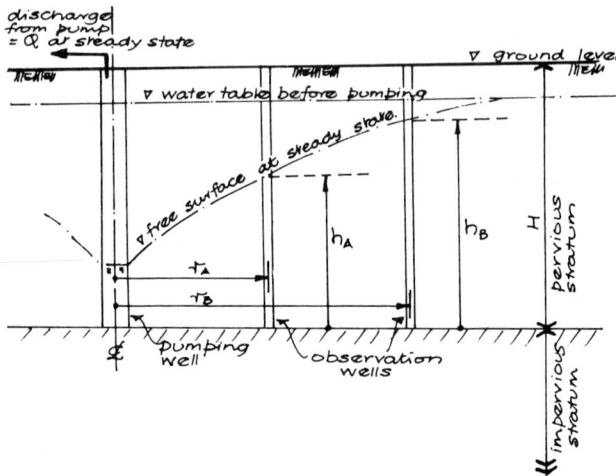

Fig. 3.6 Unconfined flow – well-pumping test

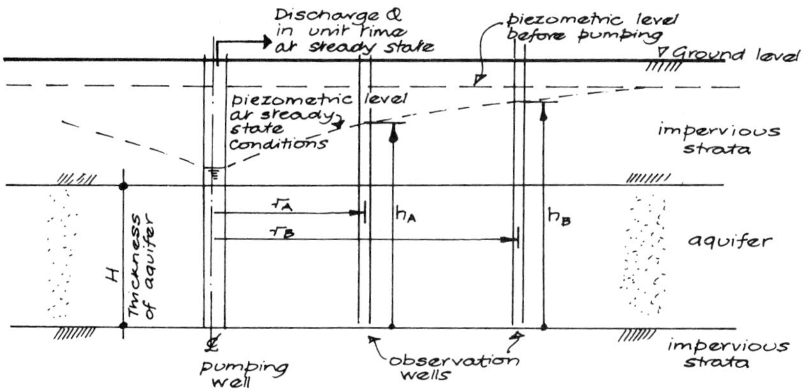

Fig. 3.7 Confined flow – well-pumping test

A similar procedure is adopted: that is, at steady state conditions measurements of pumping rate, depression of piezometric level and radius of observation wells are recorded. A coefficient of permeability for each pair of observation wells is computed and a mean value taken.

3.3.3 Summary

The following points are worthy of attention.
1. The permeability value obtained is a mean value over, perhaps, a significant plan area. There will be variations within this area and in some cases these can be considerable.
2. Difficulties arise if the initial water table or piezometric level is not horizontal. This may very well be so if natural seepage is taking place before pumping commences. Similarly, the thicknesses of the various strata may not be consistent.
3. The cost of pumping tests is likely to be large. While justified and necessary for large and important contracts, their use may not be appropriate to smaller contracts. The drilling of the wells can be expensive, as can be the pumping operations.
4. Pumping from wells and lowering the free surface can cause significant increases in effective pressure and consequently consolidation settlement to any neighbouring structures, particularly if the pumping is carried out for long periods of time.
5. The two tests described are the simplest and easiest pumping tests to perform. Soil profiles, water levels, natural variations of strata and levels are seldom, in practice, as clearly defined as is assumed in the theory on which these tests are based. Therefore the results obtained should be studied carefully and experienced judgement of the physical circumstances is required before a coefficient of permeability is reported for use in subsequent analysis.

6. Obviously, the K value found in the majority of pumping tests is that in a horizontal direction.

3.4 Determination of the Coefficient of Permeability by Means of the Rowe Consolidation Cell

A new consolidation cell was introduced in 1966 by Rowe and Barden.[4] This cell, subsequently developed by Wilkinson,[5] is admirably suited to the determination of both horizontal and vertical coefficients of permeability on large-diameter samples that are considered to be fully representative of the mass. The apparatus described here has been used successfully in a variety of tests on large-diameter undisturbed samples.

The main components of the Rowe cell are described in Chapter 4; the description here is limited to its use as a means of measuring permeability characteristics. Figure 3.8a shows the Rowe cell set-up for a vertical permeability test, while Fig. 3.8b gives the set-up for horizontal permeability observations. These set-ups are shown diagrammatically in Figs 3.9 and 3.10.

The main features are that the test is of the constant-head type and that observations of hydraulic gradient across the sample, together with the measurement of the quantity of flow through the sample, are determined more precisely than in the rather crude standard permeameter tests. A third important aspect is that the void ratio of the test sample is the same as that existing on site and the levels of total pore-water and effective pressures are likewise similar to field or anticipated field values.

3.4.1 Vertical Permeability Test

For a vertical permeability test the sample is initially consolidated to the effective pressure operative on site. This is achieved by applying hydraulic pressure to the convoluted rubber jack from an air–water cylinder. A compressor-fed air supply to the air–water cylinder is regulated to correspond to the expected total field pressure, and this regulated pressure is applied hydraulically to the sample from the air–water cylinder through the rubber jack.

The drainage outlet to the sample is then connected either to an elevated water-bottle or to a Bishop mercury pot system (see Chapter 4) set to a pressure corresponding to the *in situ* pore-water pressure. When the pore-water pressure transducer records a pressure equivalent to the pressure applied to the drainage outlet of the cell, the sample is essentially at the pressure levels and the void ratio levels required for the particular test.

Since the initial sample thickness is known, any change in thickness can be computed from observations of the dial guage attached to the drainage spindle.

The sample is initially sandwiched between two sintered bronze discs, with filter paper interposed between the bronze discs and the soil sample. It is essential that these discs should be fully saturated when the sample is set

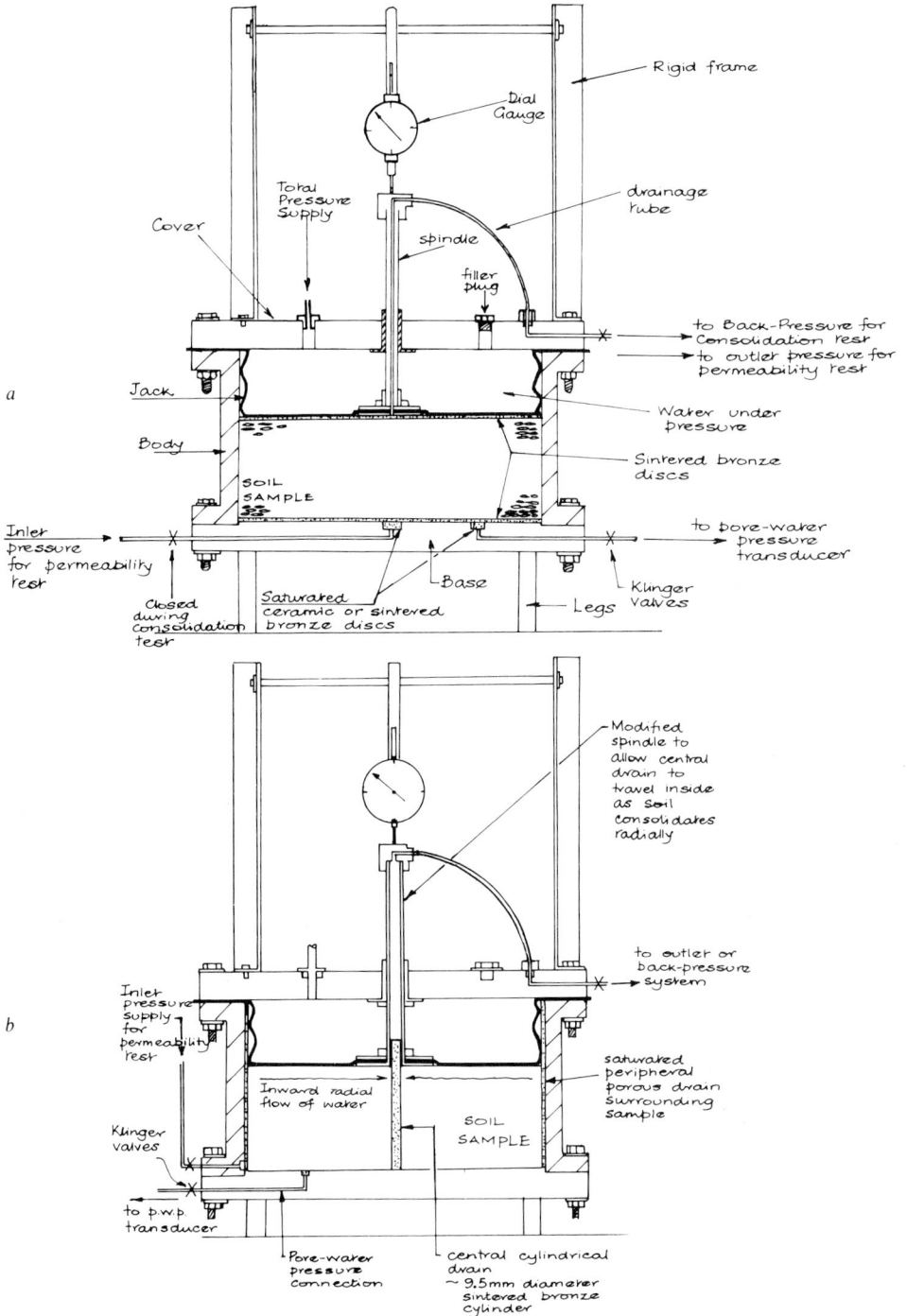

Fig. 3.8(*a*) Rowe consolidation cell, 254 mm diameter. Set-up for vertical permeability
test
  (*b*) Set-up for inward radial consolidation test and horizontal permeability test

up, and this is achieved by boiling them in water for half an hour and then allowing them to cool before use. This usually expels all air entrapped in the discs.

A Klinger valve at the base of the cell is connected to the centre of the base of the sample and to an inlet pressure supply, provided from either an elevated constant-head water bottle or a second Bishop constant-pressure mercury pot system. The outlet pressure for the permeability test is provided by the pressure supply connected to the drainage spindle. Both pressure lines are connected to a differential manometer incorporated into the circuit so that the difference between inlet and outlet pressure can be constantly monitored.

It is usual to set the inlet pressure supply (through the base of the sample) to the *in situ* pore-water pressure value and to lower slightly the outlet supply pressure (from the drainage spindle), so as to create a pressure difference across the sample. This pressure difference (recorded on the differential manometer) must be maintained at less than 10% of the effective pressure on the sample; initially it causes some minor consolidation of the sample but, again, any change in thickness (or reduction in drainage path length) is monitored by the dial gauge.

Further refinements include the filling of the whole permeability circuit with 'site' water, and the observation of the differential manometer by a travelling telescopic device.

The quantity of flow is calculated from observations taken in two flow-tubes incorporated into the permeability circuit. One flow-tube monitors 'inward' flow and the other the 'outflow' from the sample. These consist of precision-bore glass tubes (1·5 mm bore) into which a small air bubble is introduced by a screw hand-pump. As the water passes through the soil, the bubbles in the inlet and outlet lines traverse the flow tubes; by means of

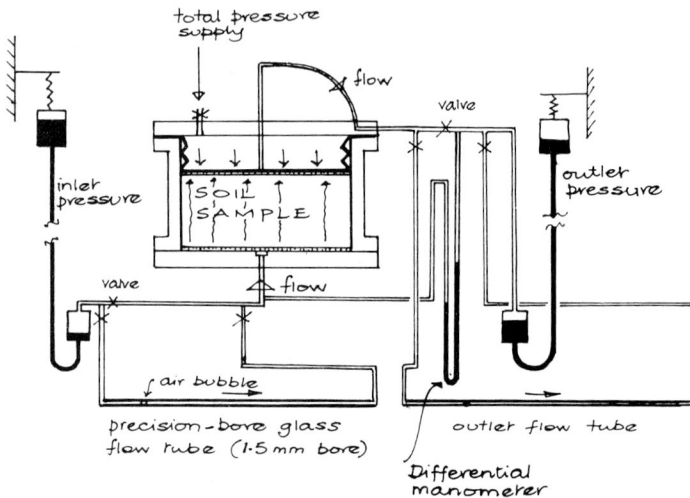

Fig. 3.9 Vertical permeability test

scales mounted alongside the tubes, the velocity of the bubbles can be evaluated. This then gives a measure of the flow quantity passing through the sample. Steady-state conditions are taken to be achieved when these rates agree within 10%. The time taken to reach steady state varies with the level of effective pressure and with soil type, from as short a period as 2 hours to up to 3 or 4 days. One essential maintenance measure is that the tubes are periodically cleansed to minimise the risk of inaccurate observations due to Jamin effects.

Typical results are given in Datasheet No. 9.

3.4.2 Horizontal Permeability Test

Determination of the coefficient of permeability in a horizontal direction may be similarly determined by using a Rowe cell. The main differences from vertical permeability tests are that the sample is surrounded on its cylindrical surface area by a drainage medium, that a central small cylindrical drain is inserted into the sample and that the inlet pressure line is connected to the cylindrical drainage medium via the cell body, rather than being connected to the base of the sample. Seepage of the water through the soil therefore takes place horizontally from the cylindrical surface of the sample to the central drain. A further important difference is that the sample diameter is less than that used for vertical permeability tests because of the need to accommodate both the outer drainage medium and the sample within the 254 mm diameter of the body.

As in the vertical permeability test, the pressure difference is monitored by a differential manometer and flow quantities are calculated from observations of air-bubble velocity in the precision-bore flow tubes.

A porous plastic circumferential drain (3·5 mm thick) has proved satisfactory. This drain is saturated by boiling in water and then cooling before use. For the central cylindrical drain two methods have been employed: (a) a central hole is carefully made in the sample with a thin-walled mandrel, and then this preformed hole is filled, under water, with a uniform graded sand; or (b) a sintered bronze cylinder is inserted into a similarly preformed central hole in the soil sample. The second method requires some modifications to the spindle arrangement of the cell to allow the rigid sintered bronze cylinder to accommodate within the spindle as settlement of the sample occurs. Figure 3.10 shows a typical set-up.

3.4.3 Summary

The Rowe cell provides perhaps the most practical means of providing valid laboratory permeability test results. The advantages may be listed as follows.

1. Tests are performed on large, representative samples which can incorporate all the natural features of the soil.
2. Field conditions can be very closely simulated. Values of void ratio are simulated and levels of total and pore-water pressures are chosen to suit site conditions.

Vertical Permeability Test using Rowe Cell

Initial thickness of sample = 75.13 mm

Initial dial gauge before consolidation = 20.00
Dial gauge at time of permeability test = 18.17
$$\text{Difference} = 1.83$$

Drainage path length = Thickness of sample = $75.13 - 1.83 = \underline{73.30mm}$
during 'k' test $= L$

Cross sectional
area of sample = $A = \frac{\pi}{4} \times 254^2 = \underline{50670.75 \text{ mm}^2}$
(254 mm diameter)

Precision flow tube data :

Test	1	2	3	4	
Inlet tube - time for 500mm travel of air bubble	473	459	465	454	sec
Outlet tube - time for 500 mm travel of air bubble	487	471	481	475	sec

mean time = 470.63 s

mean velocity = $\frac{500}{470.63}$ = 1.06 mm/s

Flow tube diameter = 1.5mm , Flow tube area = $\frac{\pi}{4} \times (1.5)^2 = 1.767 \text{ mm}^2$
mean quantity of flow = Area $\times$ mean velocity
$$Q = (1.767) \times (1.06) = \underline{1.873 \text{ mm}^3/\text{sec}}$$

Inlet pressure to base of cell = 70 kN/m² *
Outlet pressure to top drain = 65 kN/m² * $h = \frac{P_1 - P_2}{\gamma_w}$

Total pressure applied to sample = 140 kN/m² , $G_s = 2.67$
Initial moisture content (%) before consolidation = 24.7 %

* observations taken by manometer readings.

Darcy's Law : $Q = A.k.i = A.k.\frac{h}{L}$ $Q = 1.87 \text{ mm}^3/\text{sec}$
$L = 73.30 \text{ mm}$
$h = \frac{(70-65)}{\gamma_w = 10} = 0.5m = 500mm$
head of
water
$A = 50670.75 \text{ mm}^2$

$k = \frac{Q.L}{A.h} = \frac{1.87 \times 73.30}{50670.75 \times 500}$

$k = 5.4 \times 10^{-6} \text{ mm/sec}$
$= \underline{5.4 \times 10^{-5} \text{ cm/sec}}$

Void ratio at "k" test : $\frac{\Delta e}{1+e_0} = \frac{\Delta H}{H_0}$

Assuming full saturation, $e_0 = m_0 G_s = (.247)(2.67) = 0.66$

$\Delta H = 20.00 - 18.17 = 1.83 \text{ mm}$ $H_0 = 75.13 \text{ mm}$

$\therefore \Delta e = \frac{1.83 (1 + 0.66)}{75.13} = 0.04$

$\therefore e = \underline{0.62}$

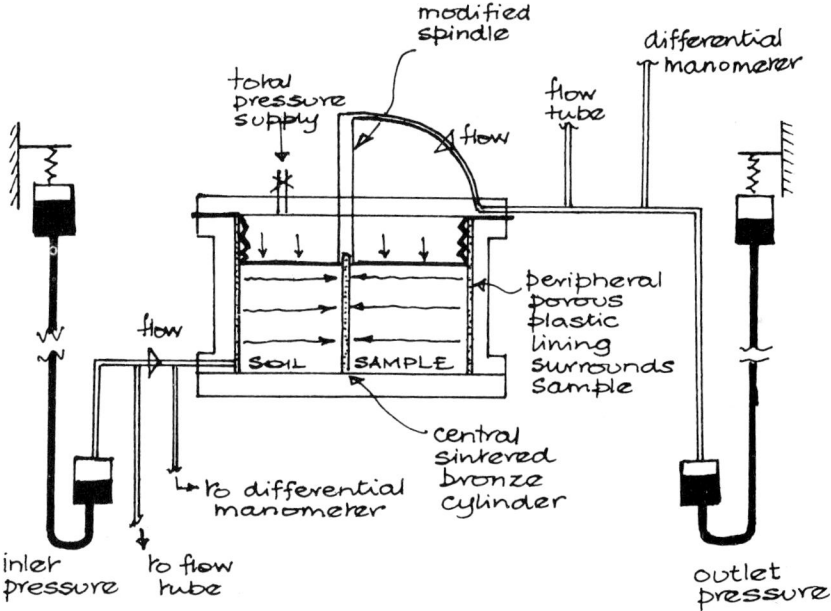

Fig. 3.10 Horizontal permeability test

3. Both vertical and horizontal permeability values can be readily determined; the only point to note is that for horizontal permeability tests
   samples of slightly smaller diameter are required.
4. If the tests are performed at various stages of consolidation tests, a
   relationship between void ratio and permeability can be established for
   a particular soil. It is widely accepted that the void ratio is linearly
   related to the logarithm of permeability.
5. The observations of seepage velocity through the soil and the hydraulic
   gradient across the sample can be measured accurately.
6. The combination of consolidation tests with permeability measurements at each increment of loading can be very conveniently undertaken. The added circuity can be easily modified to permit both horizontal and vertical measurements of permeability.

CHAPTER 4

Shear-strength Tests

Determination of the shear-strength parameters of a soil using

the unconfined compression test
the shear box test
the vane test
the triaxial test including laboratory measurement of pore-water pressures.

4.1 Introduction

The purpose of shear-strength testing of soils is twofold:

(*a*) to allow displacement under working loads to be predicted;
(*b*) to evaluate the external forces required to cause shear failure of a soil.

Testing for these purposes necessitates making certain assumptions about the behaviour of soils, the reasons for which are outlined below; discussion of the validity of these assumptions is beyond the scope of this text.

In general, soils are not elastic materials (i.e. they do not obey Hooke's law); stresses within soils are not easily related to strains by simple mathematical equations and the behaviour of soils undergoing strain is largely governed by whether the stresses are increasing or decreasing. Consequently certain simplifying assumptions are necessary for the purpose of making practical predictions, and it is important to be aware of the extent of the validity of the theory from which shear-strength test results are derived.

Other difficulties arise in that the predicted shear-strength behaviour of soils depends to a great extent upon the history of the soil, loading rates, drainage conditions, sample size, degree of saturation, boundary effects and the comparison of practical field levels of stress with the levels of stress employed during testing.

When the values of the externally applied stresses change, two possible sources of deformation of the soil arise. One is the volume change (by the expulsion of water and/or air from the soil), which results in rearrangement of the particles within the soil mass. The other is the sliding of particles relative to each other. This second effect can produce very much greater deformations than those resulting from volume change. The sliding of particles is the result of increased shear stresses within the mass and the shear strength of a soil can be said to be the resistance of a soil to such shear stresses.

4.1.1 The Mohr–Coulomb Failure Criterion for Soils

Most practical shear-strength testing of soil is based upon the Mohr–Coulomb failure criterion.

Coulomb showed that the shear strength S of a soil can be expressed by

$$S = C + (\sigma_n) \tan \varphi \tag{4.1}$$

where C and φ are shear-strength parameters and σ_n is the normal stress acting on the failure surface.

It is important to realise that C and φ are merely parameters defining the equation for shear strength as a straight line: i.e. $\tan \varphi$ = the gradient of the line, and C = the intercept on the shear-strength axis (Fig. 4.1).

Neither C nor φ is a physical property of the material, such as colour, density, odour, etc.; they can vary according to a variety of factors.

Using Terzaghi's concept of effective stress, $\sigma_n' = \sigma_n - u$, where σ_n = total normal stress, u = pore-water pressure and σ_n' = effective normal stress, Coulomb's expression (e.g. (4.1)) can be rewritten as

$$S = C' + (\sigma_n - u) \tan \varphi'$$

or

$$S = C' + (\sigma_n') \tan \varphi' \tag{4.2}$$

where C' and φ' are parameters related to effective stress.

Some problems in soil mechanics can be solved by using either total stresses (eq.(4.1)) or effective stresses (eq.(4.2)). Usually eq.(4.2) is more

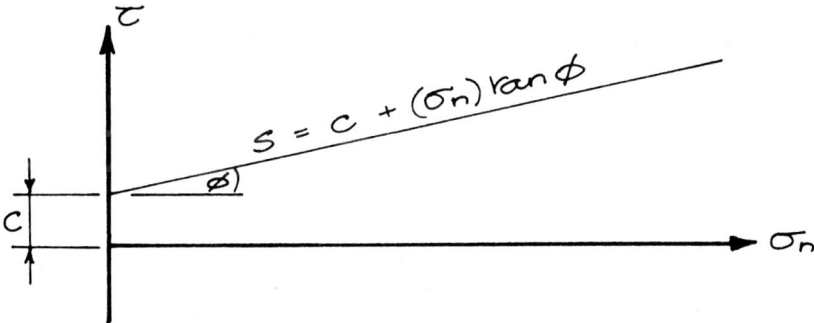

Fig. 4.1 Coulomb's equation for shear strength

relevant to practice: eq.(4.1) is only rarely employed. In fact, the choice of the relevant equation is a necessary preliminary to the setting-up of a laboratory shear-strength testing programme.

Mohr's circle construction for the graphical determination of the shear and normal stresses within an element of a material subject to externally applied major and minor principal stresses is shown in diagrammatic form in Fig. 4.2.

The envelope of failure (which is tangential to all failure circles) is generally non-linear. However, with soils, over a limited range of normal stresses, it is sensibly linear and can be combined with Coulomb's equation for the shear strength (eqs (4.1) and (4.2)) to give the Mohr–Coulomb failure criterion for soils.

If it is assumed that the Mohr envelope and the Coulomb equation are one and the same line, it follows that all combinations of σ_1 and σ_3 at failure produce Mohr's circles at failure which have a common tangent given by Coulomb's equation. It should be remembered that this is an assumption and has inherent weakness. However, for most practical purposes it describes adequately the shear strength of soils and it is readily and easily applied to the problems normally encountered.

Having determined experimentally the values of the parameters C and φ (with respect to total stress), or more usually C' and φ' (with respect to effective stress), it is then possible to estimate the shear strength of the soil by predicting values of σ_n (normal stress) and u (pore-water pressure) and substituting these values in the appropriate equation (eq. (4.1) or eq. (4.2)).

4.1.2 Test Conditions

In an attempt to simulate full-scale behaviour with regard to both loading rates and drainage conditions, three types of test are normally performed.

1. Undrained Tests

In this category of test no drainage of pore water is allowed at any stage. This implies that the sample is sheared at a constant moisture content and, if sampling, storage and setting-up of the test have been performed correctly, the moisture content during testing should correlate exactly with the moisture content of the soil in its natural state at the time of sampling on site. It also implies that if the soil is *fully* saturated then, with no drainage during testing, there is therefore no volume change during testing.

2. Consolidated Undrained Tests

In consolidated undrained tests the sample is initially allowed to consolidate under an effective stress corresponding to the effective stress *in situ* at the site, i.e. the moisture content is reduced from its initial value. Consolidation is the process whereby pore water is expelled from a saturated soil under constant total stress. Thus drainage is allowed, the volume of the sample decreases and completion of consolidation is marked by a cessation of fur-

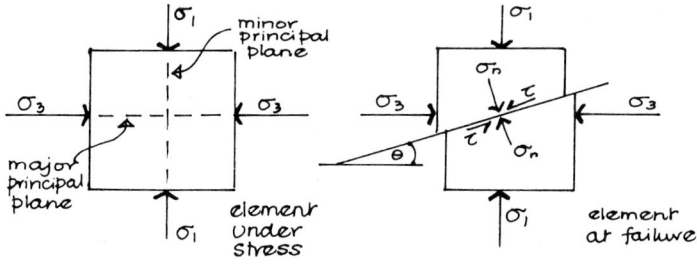

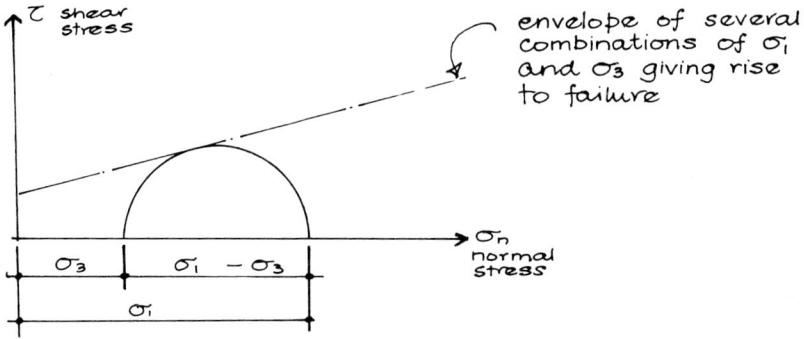

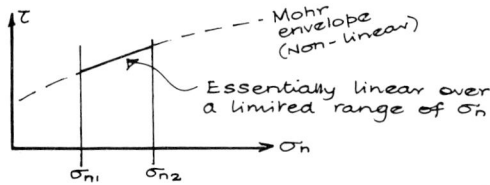

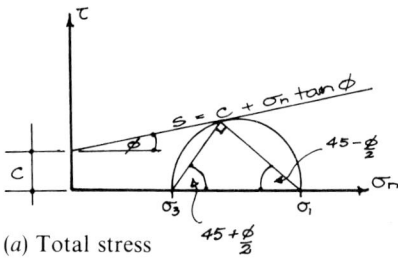

(a) Total stress   (b) Effective stress

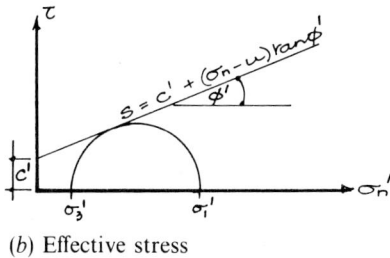

Fig. 4.2 Mohr's circle construction

σ_1 = major principal stress
σ_3 = minor principal stress
σ_{n} = normal stress
τ = shear stress
θ = inclination of failure plane to major principal plane
Note: $\sigma_3{}' = \sigma_3 - u; \sigma_1{}' = \sigma_1 - u$

ther volume change or expulsion of water. The time required to complete consolidation obviously varies from soil to soil with varying permeability.

When consolidation is complete, the sample is sheared with *no drainage* (i.e. undrained) and, therefore, shearing is carried out at the moisture content reached at the end of consolidation. No volume change is recorded or allowed during this shearing process.

3. Drained Tests

The drained test is performed under conditions whereby drainage of the sample is allowed at all times. Consequently there is a continual reduction in moisture content of the soil from that obtaining at the time of sampling. Associated with this continually changing moisture content are, obviously, continual volume changes of the sample during both consolidation and shearing stages of the test. Another condition of the drained test is that the loading is applied so gradually that there is no or very little build-up of pore pressure. Each increment of load is applied only when the pore-water pressure within the sample has fallen to the value existing before the application of the previous load increment. The significance of this will be illustrated later.

4.1.3 Introduction to Stress Path Concept

When performing 'sets of three' triaxial tests and then plotting the corresponding Mohr circle diagrams, there can be some discrepancy in drawing the envelope as the best common tangent and the judgement involved in choosing this tangent can often vary significantly from person to person.

In addition when such 'sets of three' tests are performed over a fairly wide stress range, then a variety of C and φ values can be obtained due to a quite natural scatter of the experimental data, possibly indicating wide variation in the value of the shear strength parameters.

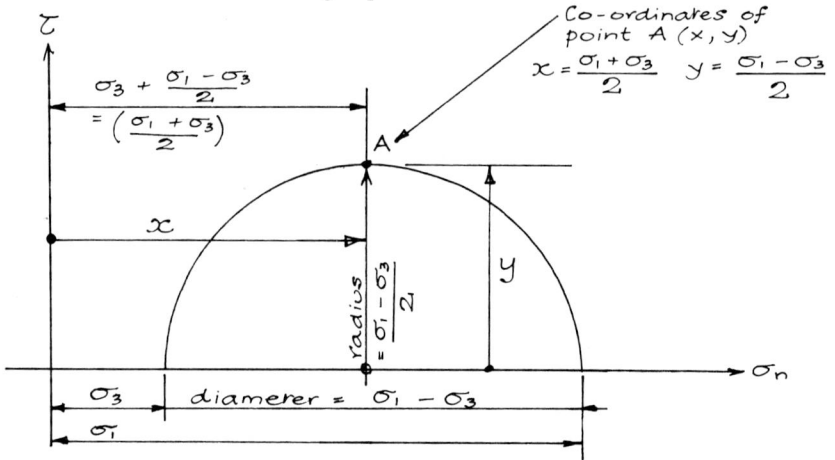

Fig. 4.3

An alternative mode of plotting the results, which also serves as an introduction to stress path concepts, is to plot the topmost of the Mohr circle diagram as indicated in Fig. 4.3. Thus for several such Mohr's circles at failure, by plotting the topmost point of each circle (such as A) with values of $(\sigma_1 + \sigma_3)/2$ as abscissa and $(\sigma_1 - \sigma_3)/2$ as ordinate the result will be as shown in Fig. 4.4. This then encourages the drawing of a line that fits the

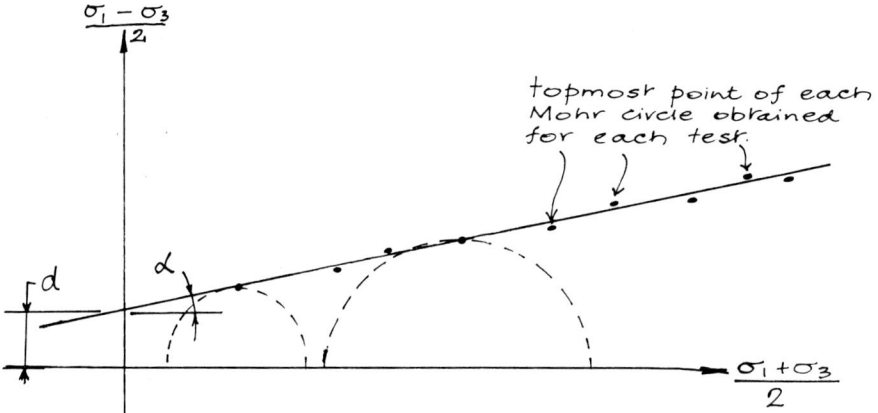

Fig. 4.4

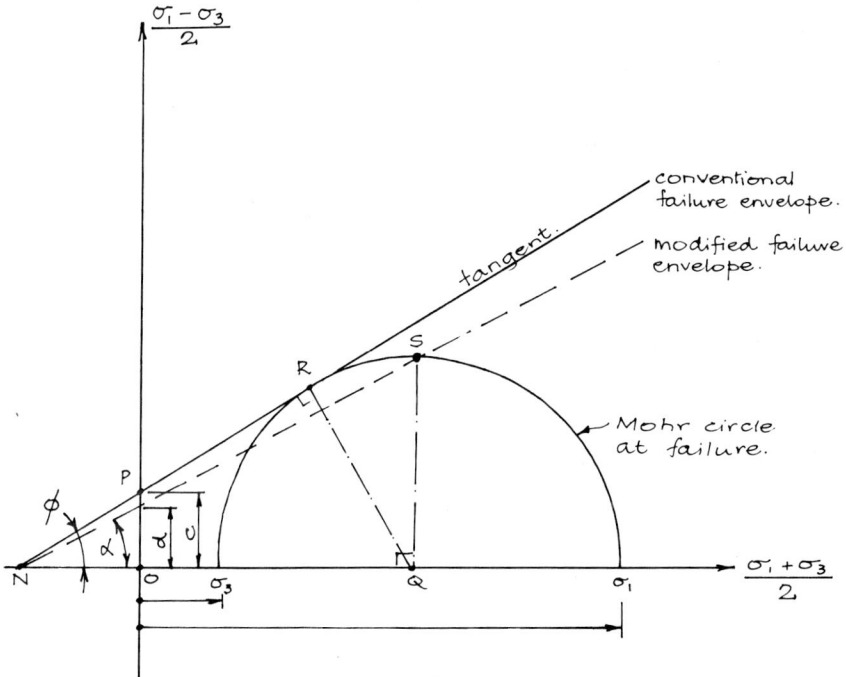

Fig. 4.5

experimental data with little variation of judgement as to which is the 'best' line. In fact this line can be readily determined using a simple statistical technique. By using this technique one eliminates the need to draw a circle for each test since it is only necessary to plot the topmost point of each Mohr circle.

Conversion of the gradient of the line (tan α) and the intercept (d) to the parameter values of 'C' and 'φ' is then carried out by fairly simple trigonometry.

Considering Fig. 4.5

$$\tan \varphi = \frac{C}{NO} \text{ and } \tan \alpha = \frac{d}{NO}$$

or

$$NO = \frac{C}{\tan \varphi} = \frac{d}{\tan \alpha} \text{ or } C \cot \varphi = d \cot \alpha$$

From triangles NRQ and NSQ

$$C\left(\frac{NR}{RQ}\right) = d\left(\frac{NQ}{SQ}\right)$$

Since RQ = SQ = radius of circle

then $$C(NR) = d(NQ)$$

$$C = d\left(\frac{NQ}{NR}\right) \text{ where } \frac{NR}{NQ} = \cos \varphi$$

thus $$C = \frac{d}{\cos \varphi}$$

From triangle NRQ $$\sin \varphi = \frac{RQ}{NQ}$$

and from triangle NSQ $$\tan \alpha = \frac{SQ}{NQ}$$

and since RQ and SQ are both radii of the same circle, it follows that

$$\tan \alpha = \sin \varphi$$
or alternatively $$\varphi = \sin^{-1}(\tan \alpha)$$

By plotting the results of shear strength tests in this way, several important features are highlighted. These include:

(i)   The elimination of errors of judgement in choosing the best common tangent.

(ii) The avoidance of apparently ignoring the weakest specimens in the test series (see Fig. 4.6).

(iii) The continual monitoring of $(\sigma_1 + \sigma_3)/2$ and $(\sigma_1 - \sigma_3)/2$ values during the shear strength tests then gives a 'stress path' showing the successive states of stress within a sample as it approaches failure. Depending upon the types of test, soil and stress path chosen these lines can be curves or linear as illustrated in Fig. 4.7.

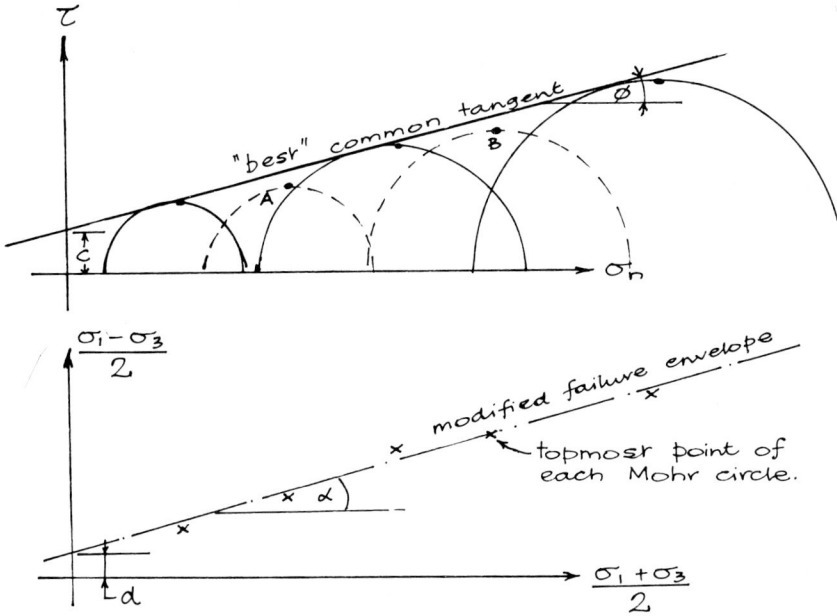

Fig. 4.6

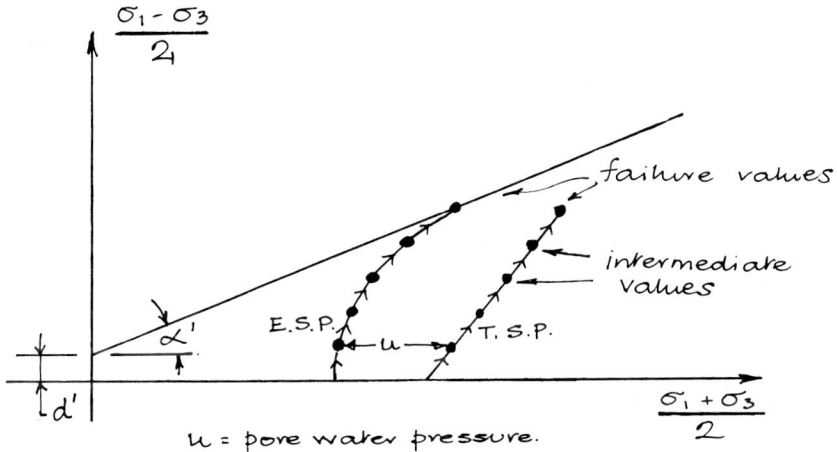

Fig. 4.7 Typical stress path – consolidated undrained test on an overconsolidated clay
Note: The horizontal difference between the T.S.P. and E.S.P. at any stage of the test will represent the pore-water pressure (u) at the stage of the test

Stress paths can be plotted as

a total stress path (T.S.P.) or effective stress path (E.S.P.)

giving appropriate failure parameters of C and φ or C' and φ'.
For a normally consolidated clay undergoing a similar consolidated undrained test the resulting stress paths will be as shown in Fig. 4.8.

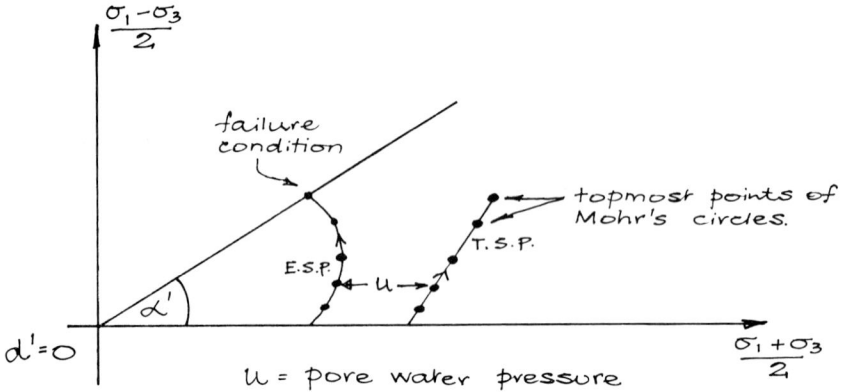

Fig. 4.8

4.2 The Unconfined Compression Test

The essential features of the unconfined compression test apparatus are
shown in Fig. 4.9. The sample is fitted between two conical platens, the
upper platen being fixed while the lower platen is capable of upward vertical
movement. A handle is connected to a linear spring and movement of the
handle raises the backing plate which, in turn, raises the lower platen in a
vertical direction, causing compression of the sample. As the handle is
rotated, the linear spring is extended and a pivoted pencil-arm attached to
the lower platen traces the movement of the platen on the backing plate.

A variety of springs is available and each spring has a calibration tracing
mask, which, when placed over the record of the pencil movement, relates
movement of the spring (i.e. movement of the sample) to the vertical force
exerted by the spring. As the cross-sectional area of the sample is known,
this force can readily be converted to a corresponding vertical compressive
stress.

4.2.1 Procedure

Assuming that either remoulded or undisturbed samples are available in 38
mm diameter thin-walled sampling tubes, the following procedure can be
adopted.

The sample is extruded carefully and the ends coned by the available
coning tool. Usually they are coned so that the sample is 38 mm in diameter
by 76 mm long (Fig. 4.10). The sample is also weighed to an accuracy of 0·5 g
so that a bulk unit weight determination can be made. In addition, three

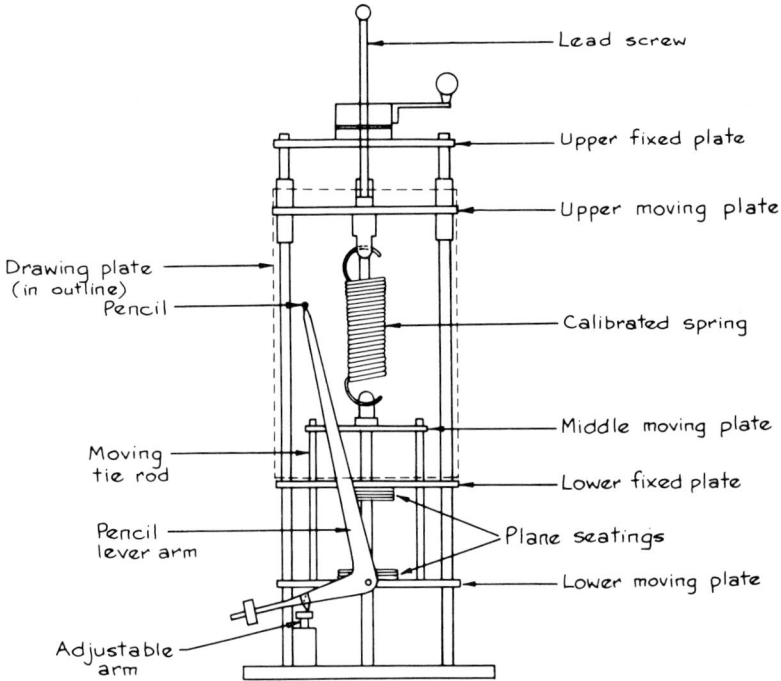

Fig. 4.9 Unconfined compression apparatus

samples of the cuttings from the sample tube should be taken, and used to determine average moisture content.

The sample is placed on the lower moveable platen and the handle is rotated until the sample just seats into the fixed top platen. The pivot arm and pencil are adjusted on the backing plate to allow the chart and pencil to coincide on the zero deflection line. The zero load position should be clearly marked.

The handle is then rotated to produce a compressive strain rate of about 10% per minute (i.e. approximately 8 mm per minute compressive deformation of the sample).

One of two types of failure may occur: either a brittle-type failure with evidence of shear planes developing, or a plastic bulging failure. Each type of failure usually displays a distinctive pencil record on the backing sheet. The two types of failure and corresponding traces are shown in Fig. 4.11.

With the plastic failure there is no clear reduction of load at failure and the failure load is taken as the load at 20% strain (i.e. 15·2 mm compressive deformation of a 76 mm sample). The brittle-type failure produces a distinct maximum compressive load clearly defining failure.

The corresponding compressive failure stress is then calculated assuming

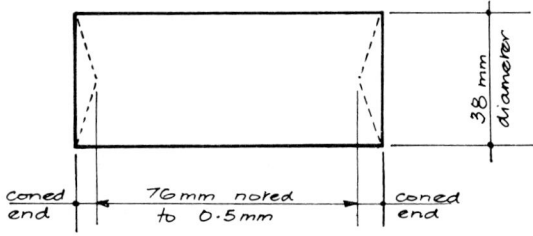

Fig. 4.10

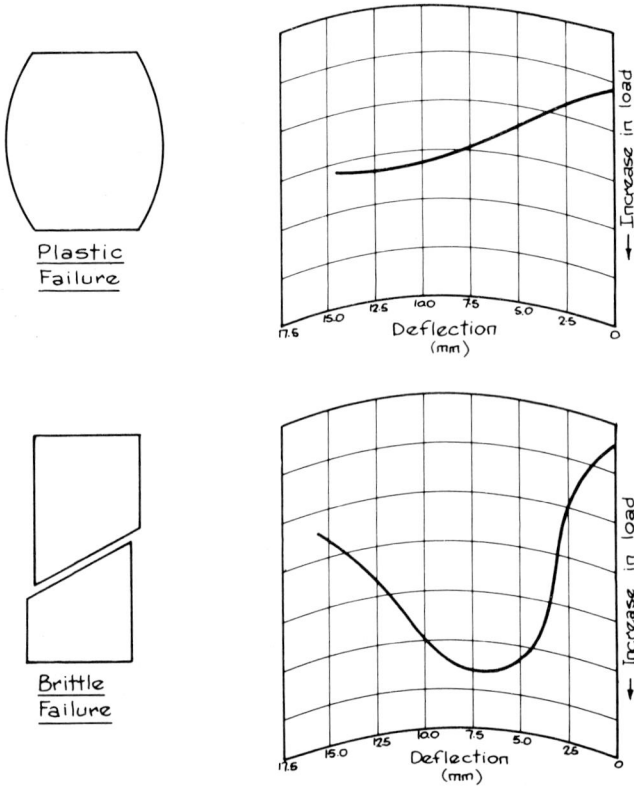

Fig. 4.11 Typical results of unconfined compression test

constant volume of the sample, i.e. vertical compressive deformation results in an increase in the cross-sectional area of the sample.

4.2.2 Summary

1. In the unconfined compression test the minor principal stress σ_3 is zero so that the corresponding Mohr's circle always passes through the origin when total stresses are plotted. The resulting failure envelope is horizontal and the corresponding shear-strength parameters defining this horizontal line are limited to an intercept value C_u, φ_u being zero because the envelope line is horizontal.
2. The type of sample that can be tested is limited by the lack of a lateral confining pressure.
3. The conditions of stress and levels of pore-water pressure are of interest and are discussed later when considering the triaxial test.

4.3 The Shear Box Test

The shear box test is often referred to as the direct shear test because an attempt is made to relate shear stress at failure directly to normal stress, thus directly defining the Mohr–Coulomb failure envelope.

Essentially, a sample of soil is subject to a fixed normal stress and a shear stress is induced along a predetermined plane until shear failure of the soil takes place. The apparatus is shown diagrammatically in Fig. 4.12. The soil sample is usually square in plan and rectangular in cross-section. Shear boxes are available in plan sizes from 64 mm square up to 254 mm square. The sample can either be remoulded or undisturbed and the box is capable of accepting coarse-grained soils. Obviously, when testing remoulded or coarse-grained samples care must be taken to ensure that the tested samples are prepared at bulk unit weight and moisture content values relevant to the problem under consideration. This can be difficult to achieve.

The box is of rigid metal construction, open at the top, and is immersed in a water-container. The box is manufactured in two halves so that the upper half is able to move horizontally relative to the lower half; the sample can thus be sheared on a horizontal plane.

The normal load is applied vertically to the sample, usually through a loading platen by means of hanger weights. For higher normal stresses a 5:1 or 10:1 lever mechanism can be used.

The shear force is applied horizontally to the sample by a motor-driven push rod. The motor drive is usually multi-geared so that a variety of shear loading rates can be applied. To facilitate free movement of the lower and upper halves of the shear box, the box is mounted on ball-bearing slides.

The shear load applied to the sample is recorded by means of a proving-ring mounted in a horizontal plane. Deformation of the proving-ring, monitored by a dial gauge, is related to the shear load applied by means of a calibration graph.

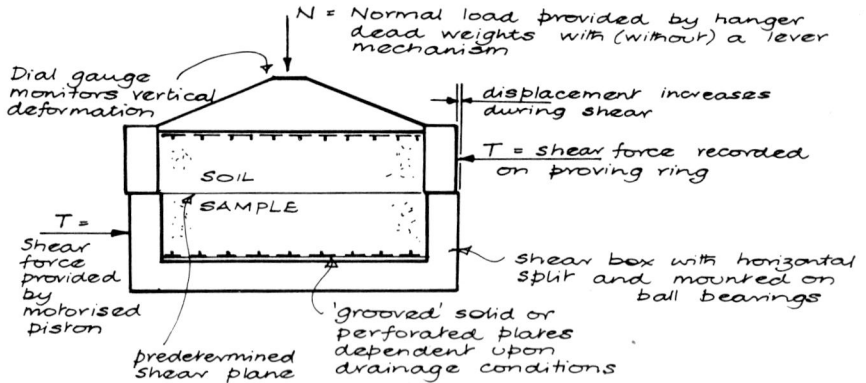

Fig. 4.12 Shear box test

In addition, it is usual to monitor vertical deformation of the sample by a dial gauge fixed to the top loading platen.

Tests can be performed under either undrained or drained conditions by the insertion of either solid metal or perforated metal plates adjacent to the soil sample's upper and lower faces. Usually these metal plates are grooved to facilitate 'grip' on the faces of the sample. A typical set-up is shown in Fig. 4.13.

4.3.1 Procedure

With fine-grained soils in either a remoulded or undisturbed state, the first step is to trim the sample carefully to suit the plan dimensions of the box, and to take moisture content samples from the resultant cuttings. A manufactured trimmer is usually available to suit the particular shear box being used. With coarse-grained soils the sample preparation is more difficult and requires great care and expertise to attain the required sample condition. The larger boxes are more appropriate and careful tamping in layers is required, together with relevant weighing operations to determine bulk unit weight values.

Assuming that the drainage conditions of the test have been predetermined, the corresponding metal plates are inserted into the shear box adjacent to the upper and lower faces of the sample.

The hanger weight applying the vertical normal stress to the sample is then assembled, together with the dial gauge monitoring vertical displacements. Locating screws which align the two split halves of the box are removed and an appropriate gear is selected for the required rate of application of the horizontal shear force. Shearing of the sample is then begun and records of time, proving-ring dial gauge and vertical-deformation dial gauge are kept throughout the test until shear failure of the sample takes place. The point of failure is signified by a fall-off in recorded shear load (or proving-ring dial gauge) with continued separation of the two halves of the shear box.

The whole test procedure is then repeated with at least three similar samples of the same soil, each with an increased normal load applied during horizontal shearing of the sample.

Typical results are given in Datasheet No. 10.

4.3.2 Summary

1. There is some measure of control of drainage during testing, and both undrained and drained tests can be performed. The drained tests must be carried out at such a low rate of shearing that no excess pore-water pressures are allowed to develop.
2. During testing there is no available means of measuring the pore-water pressure. In some particular field problems a knowledge of pore-water pressure during shear is a distinct advantage and will aid field predictions of shear strength.
3. Other drawbacks to the shear box test are that the plane of shear is predetermined as horizontal by the split box; the area of shearing is constantly decreasing as the two halves of the soil separate; the results can be affected by end and side effects of the rigid box and there is subsequent doubt about the stress distributions within the sample.
4. On the other hand, in some cases the shear box can be of distinct

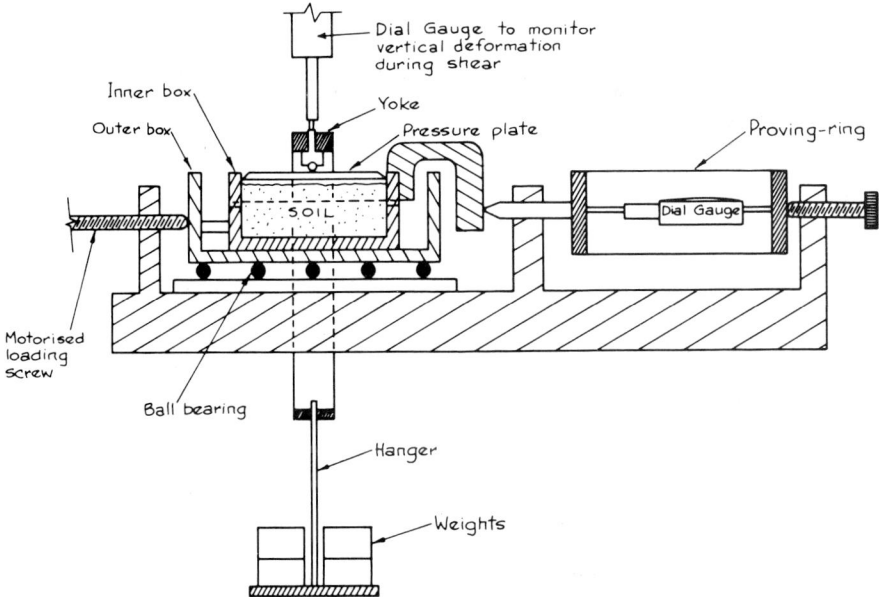

Fig. 4.13 Shear box apparatus (not to scale)

Shear Box Test (Undrained)

Time min	Strain %	TEST 1 Normal stress = 200 kN/m²				TEST 2 Normal stress = 300 kN/m²				TEST 3 Normal stress = 400 kN/m²			
		Proving Ring Dial Gauge Divs	Shear Force kN	Shear Stress kN/m²	Vertical Deformation Dial Gauge Divs	Proving Ring Dial Gauge Divs	Shear Force kN	Shear Stress kN/m²	Vertical Deformation Dial Gauge Divs	Proving Ring Dial Gauge Divs	Shear Force kN	Shear Stress kN/m²	Vertical Deformation Dial Gauge Divs
0	0	0	0.0	0.0	20.00	0	0.0	0.0	20.00	0	0.0	0.0	20.00
¼	0.5	31	0.031	7.7	19.97	61	0.061	15.1	19.94	101	0.101	25.1	19.98
½	1	52	0.052	12.9	19.96	118	0.118	29.2	19.92	217	0.217	52.5	19.92
¾	1.5	81	0.081	20.0	19.91	168	0.168	41.7	19.89	306	0.306	75.9	19.90
1	2	117	0.117	29.3	19.90	222	0.222	55.0	19.87	383	0.383	95.0	19.87
1¼	2.5	144	0.144	35.8	19.90	265	0.265	65.6	19.85	459	0.459	113.8	19.83
1½	3	185	0.185	45.9	19.89	303	0.303	75.1	19.85	524	0.524	130.0	19.79
1¾	3.5	222	0.222	55.0	19.88	348	0.348	86.2	19.85	588	0.588	145.9	19.78
2	4	249	0.249	61.7	19.87	383	0.383	95.0	19.84	633	0.633	157.0	19.74
2¼	4.5	283	0.283	70.0	19.86	421	0.421	104.3	19.82	670	0.670	166.1	19.73
2½	5	305	0.305	75.7	19.85	460	0.460	114.1	19.80	702	0.702	174.2	19.71
2¾	5.5	328	0.328	81.3	19.84	495	0.495	122.7	19.80	726	0.762	180.0	19.69
3	6	353	0.353	87.5	19.83	524	0.524	130.0	19.78	754	0.754	187.1	19.67
3½	7	386	0.386	95.7	19.82	565	0.565	140.0	19.76	799	0.799	198.2	19.64
4	8	403	0.403	100.0	19.82	605	0.605	150.0	19.74	821	0.821	203.6	19.62
4½	9	407	0.407	101.0	19.81	613	0.613	151.9	19.72	826	0.826	204.8	19.60
5	10	406	0.406	100.9	19.81	612	0.612	151.7	19.72	826	0.826	204.8	19.60

Area of Box = 63.5 mm × 63.5 mm = 4032.25 mm² = 4.033 × 10⁻³ m²	
Proving Ring : from calibration graph : 1 div = .001 kN	
Length of Box = 63.5 mm	Thickness of sample = 16.04 mm
Natural moisture content = 14.7 %	Bulk density = 1.831 Mg/m³

SHEAR BOX TEST (UNDRAINED)

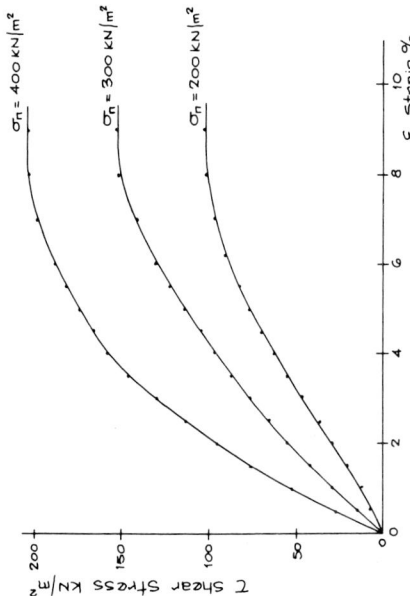

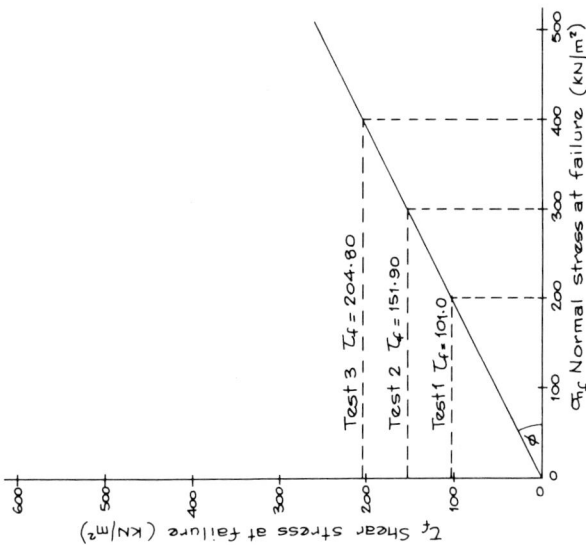

$\text{Test 3}\ \tau_f = 204.80$

$\text{Test 2}\ \tau_f = 151.90$

$\text{Test 1}\ \tau_f = 101.0$

$\tan \phi = \dfrac{256}{500} = 0.512$

$\phi = 27.1°$ $c = 0\ \text{kN}/\text{m}^2$

σ_f Normal stress at failure (kN/m^2)

τ_f Shear stress at failure (kN/m^2)

$\sigma_n = 400\ \text{kN}/\text{m}^2$

$\sigma_n = 300\ \text{kN}/\text{m}^2$

$\sigma_n = 200\ \text{kN}/\text{m}^2$

ε Strain %

τ Shear Stress kN/m^2

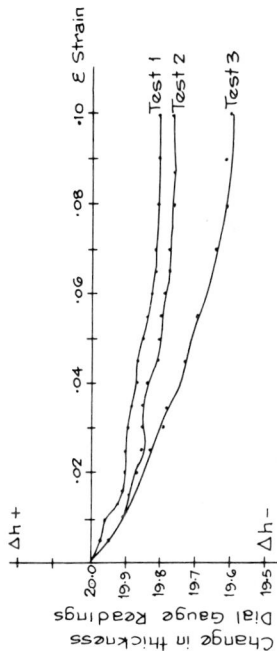

ε Strain

Test 1

Test 2

Test 3

$\Delta h +$

$\Delta h -$

Change in thickness
Dial Gauge Readings

advantage. With coarse-grained soils, provided that densities can be simulated, it can provide a relatively cheap means of estimating drained shear strength parameters. This is mainly because of the relative expense and difficulty of preparing samples of coarse-grained soils for other methods of shear-strength testing.

5. The shear box can also be used to determine residual shear-strength values (i.e. shear-strength parameters at large displacements) by reversing the travel of the machine. It should be noted, however, that a new form of shear box has been developed for such residual-strength value determinations.[6] The ring shear box, as this is called, essentially shears an annular sample of soil which is laterally confined and subject to a constant normal load. In this way no reversal is required and, with convenient gearing, large strains can be developed at slow rates of shearing.

4.4 The Vane Test

The vane test gives a rapid estimation of the undrained shear strength of fine-grained soils and can be of use in certain field situations. The vane is available in two forms, one appropriate for laboratory testing and the other for field or *in situ* testing. The laboratory vane, which is little used in practice, is essentially a small-scale version of the field vane. The field vane test

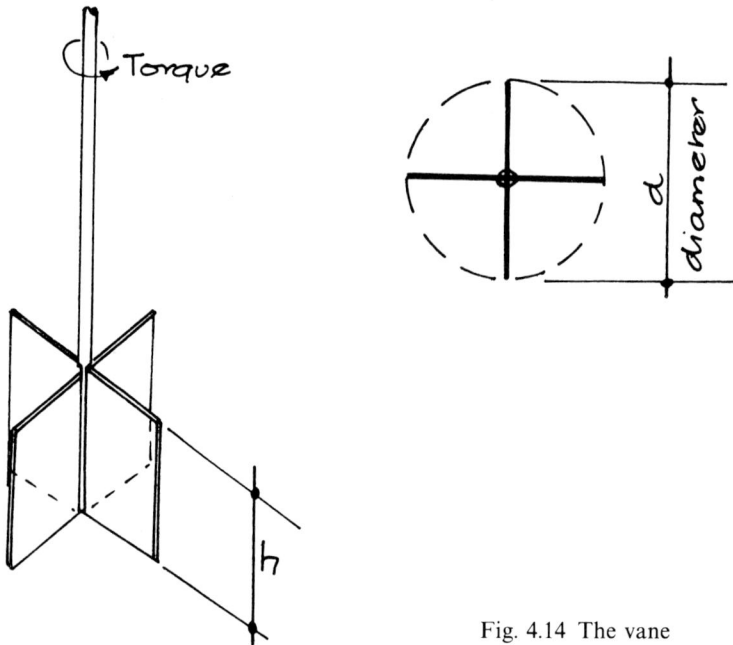

Fig. 4.14 The vane

can be performed in virgin ground or in the base of a trial pit or borehole.

Both types of vane consist of four blades set at right angles, attached to a central circular rod (Fig. 4.14). The vane is pushed into the soil until it is totally embedded and a torque is applied to the circular rod and is steadily increased. Failure of the soil is noted by a reduction in the torque required; it is assumed that the soil fails on a cylindrical surface area enclosing the extremities of the four blades (Fig. 4.15).

The physical dimensions of the vane are recorded, together with the maximum torque achieved during shear. In the laboratory, maximum torque is usually calculated from a record of the maximum horizontal deflection angle of a spring attached to the circular rod; in *in situ* testing a torque wrench or similar device records the failure value.

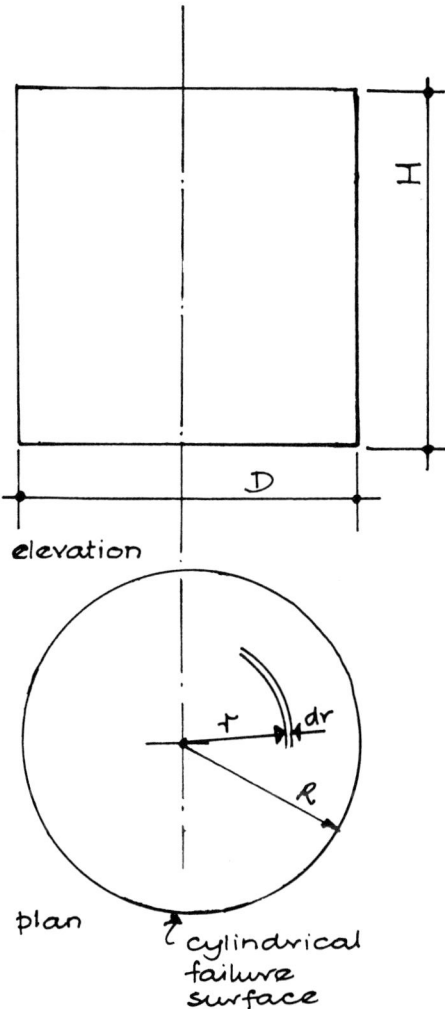

Fig. 4.15 Vane test

Assuming that the distribution of shear strength around the cylindrical surface area and around the circular ends is uniform, the torque T at failure is given by the following expression, where C_u is the undrained shear strength, H is the height of vane and D is diameter of vane:

$$T = C_u \times \text{cylindrical surface area} \times \text{lever arm}$$

$$= C_u \times \int_0^R 2\pi r \, . \, dr \times r \times (2 \text{ No. ends})$$

$$= C_u \pi DH \times \frac{D}{2} + 4\pi C_u \left(\frac{r^3}{3}\right)_0^R$$

$$= C_u \left[\frac{\pi D^2 H}{2} + \frac{4\pi}{3}(R^3)\right]$$

$$= C_u \left(\frac{\pi D^2 H}{2} + \frac{\pi D^3}{6}\right)$$

$$= \pi D^2 C_u \left(\frac{H}{2} + \frac{D}{6}\right)$$

and

$$C_u = \frac{T}{\pi D^2 \left(\dfrac{H}{2} + \dfrac{D}{6}\right)}$$

4.4.1 Summary of Vane Test

1. The vane test is particularly useful in soft, sensitive soils where there may be excessive disturbance on sampling or, indeed, where sampling cannot conveniently be undertaken.
2. The laboratory vane is little used but the *in situ* vane test has wide practical applications and some points of interest with respect to field testing are listed below.
3. The vane blades are usually manufactured from high-quality stainless steel and usually have a height-to-diameter ratio of 2:1. Common sizes are 150 mm long × 75 mm diameter and 100 mm long × 50 mm diameter.
4. The blades should be thin and should have a cutting edge on their lower edge. The area ratio is usually less than 12% and is given by the expression

$$\text{Area ratio (\%)} = \frac{8t(D - d) + \pi d^2}{\pi D^2} \times 100$$

where t = thickness of blade (mm)
d = diameter of vane rod (mm)
D = overall diameter of blades when rotated (mm).

The purpose of these conditions is that both remoulding and disturbance effects are kept sensibly minimal.
5. The test should be performed with the top of the vane blades at least 0·5
   m below the base of the borehole or trial pit so as to ensure that the soil
   being tested has not been disturbed significantly by boring or excavation operations.
6. To maintain verticality down the borehole, steady bearings may need
   to be fitted to the vane rods. Some vanes are also fitted with a protective shoe from which the vane is pushed when it has penetrated to the
   required depth – this is particularly useful when the vane is introduced
   directly into the soil from ground level.
7. The angular rate of rotation is usually of the order of 9° per minute.

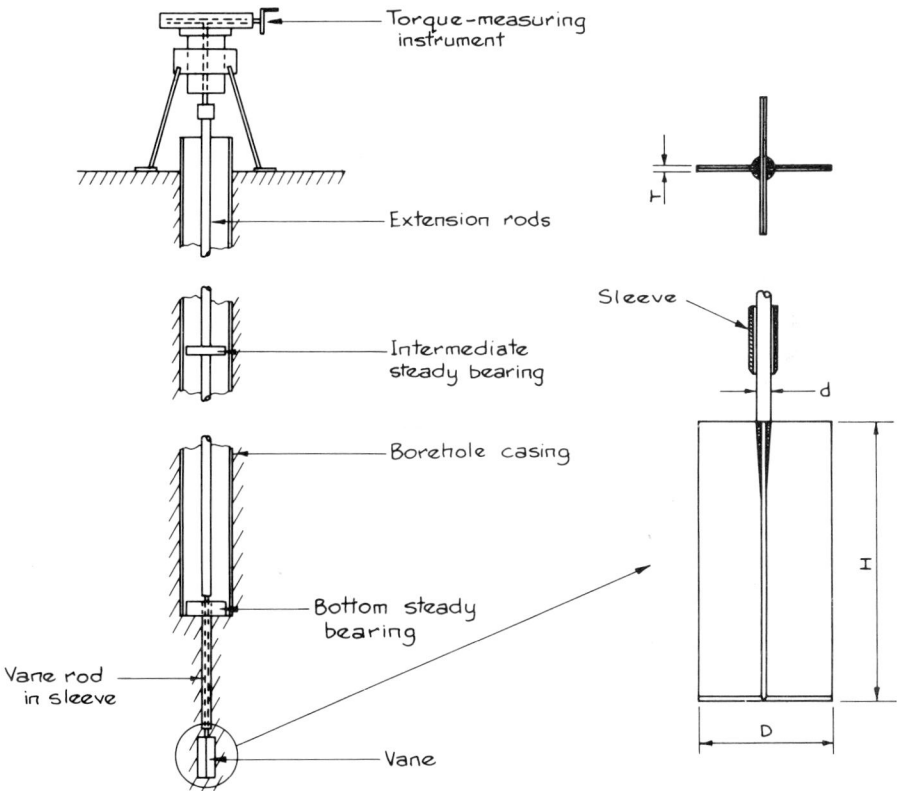

Fig. 4.16 General arrangements for the field vane test

4.5 The Triaxial Test, Including Laboratory Measurement of Pore-water Pressure

The triaxial apparatus is possibly the most widely used and most versatile means of observing the shear-strength characteristics of soils. A cylindrical sample of soil is enclosed in a pressurised chamber which subjects the sample to compressive stresses in three mutually perpendicular directions. The vertical compressive stress is then increased in excess of the horizontal stresses until eventually the soil fails in shear or strains to such a point that excessive deformation results. Means are provided to vary the drainage conditions, to monitor vertical deformations, to observe volume change of the sample and to monitor pore-water pressures during testing.

The various component parts and facilities are described with reference to Fig. 4.17 and a complete assembly is shown in Figs 4.17, 4.20, 4.22 and 4.27. Typical triaxial test results are given in Datasheets Nos 11–13 (pages 105–12, 114).

4.5.1 The Basic Triaxial Cell and Ancillary Systems

As shown in Fig. 4.17, the soil sample is mounted on a pedestal base which is attached to the base plate of the triaxial cell. The cell is formed from a Perspex cylinder attached, usually by rods, to metal upper and lower plates. The whole assembly is able to withstand, without leakage, hydraulic pressures.

The pedestal base is such that it will receive cylindrical soil samples and the standard diameters are 38 mm and 100 mm, although some cells are manufactured to accommodate samples of 254 mm diameter. The heights of the samples are usually such as to give a height-to-diameter ratio of 2:1 (i.e. 76 mm and 200 mm); as will be discussed later, modifications to the set-up can allow satisfactory testing of the larger-diameter samples with height-to-diameter ratios of unity.

The sample is enclosed in a thin rubber membrane so that it is effectively sealed from interaction with the fluid used to provide the all-round (cell) pressure. Water is the fluid most commonly used to provide this all-round pressure although recent research has shown that there can be benefit from using other fluids, such as light oil, especially in long-term tests. The function of the surrounding membrane is to prevent moisture movement across it; any potential moisture seepage between the membrane and the upper platen and base pedestal is prevented by the clamping effect of two or more 'O'-ring seals.

From the base pedestal a connection, via a Klinger valve (A), connects the pore water of the soil to a pore-water pressure measuring device; from the top platen (usually made of Perspex) is a connection via a Klinger valve (B) to a burette system to allow drainage and volume-change observations. A further Klinger valve (C) is connected to the base plate of the cell, allowing the fluid in the cell and surrounding the externally sealed sample to be pressurised.

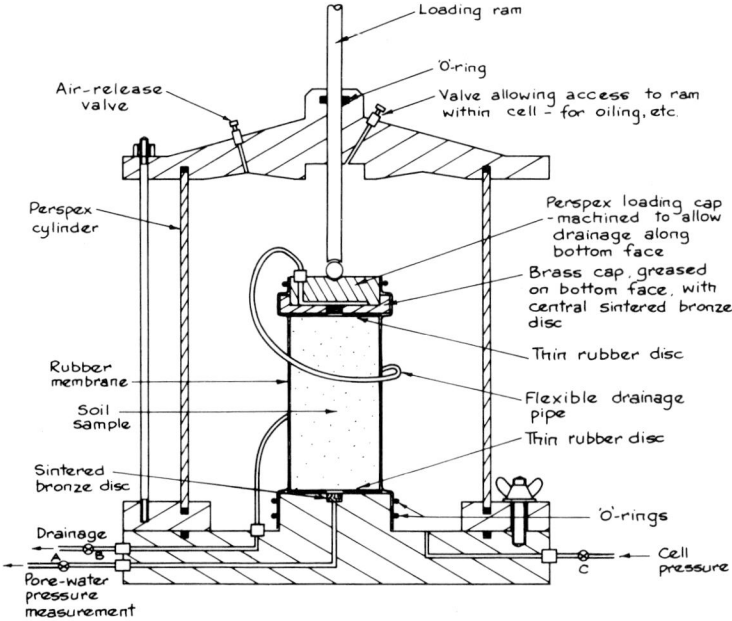

Fig. 4.17 Triaxial cell set-up

A ball seating at the upper face of the top Perspex platen allows the vertical axial load to come into contact with the sample by means of a plunger fitted in a rotating bush in the top plate of the cell.

Application of Cell Pressure (Fig. 4.18)
The two most common means of supplying a constant fluid pressure to surround the sample are

(i) a constant-pressure mercury pot system
(ii) an air–water cylinder.

The most important requirements are that the pressure supply is maintained constant for the duration of the test (which may be up to or over one month) and that the supply system can accommodate the expected volume change of the sample. In addition it is necessary to have a connection from the cell pressure supply lines to a pressure gauge or, preferably, a manometer so that cell pressures can be adequately monitored. It should be noted that the datum for pressure observations is usually taken as the mid-height of the sample.

Application of Vertical Load
The vertical load, transmitted to the sample via the plunger fitted into the top plate of the triaxial cell, is increased gradually to cause shear failure of the sample. Two systems of loading are generally available. The most common load system is one using a motorised load frame where the cell base is

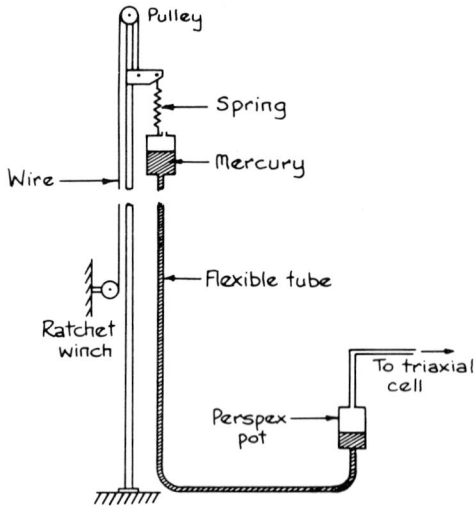

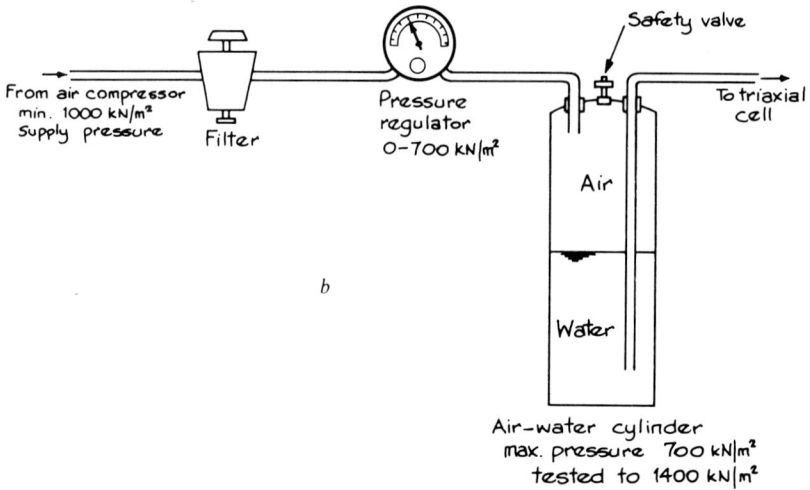

Fig. 4.18(a) Bishop constant-pressure mercury pot system
(b) Air–water cylinder, max. pressure 700 kN/m² tested to 1400 kN/m²

raised, pushing the sample and plunger against a proving-ring fixed to the load frame. Figure 4.19 shows such an arrangement.

The rate of loading can be varied by appropriate gear selection to give a variety of compressive strain rates. In undrained triaxial tests the rate of strain is often 2% per minute, thus producing 20% strain or failure within 10 minutes from the start of loading; in drained tests the strain rate may be of the order of 0·001% per minute, giving a time to 20% strain of about 2 weeks. The actual rate of strain in a drained test should be chosen so that no excess pore-water pressure builds up. It is therefore largely governed by the ability of the soil under investigation to dissipate such excess pore-water pressure.

Tests where the load application is by a motorised loading frame are termed 'strain-controlled' tests, since the vertical compressive strain is set to a fixed rate and the vertical compressive stress is such as to allow the fixed strain rate.

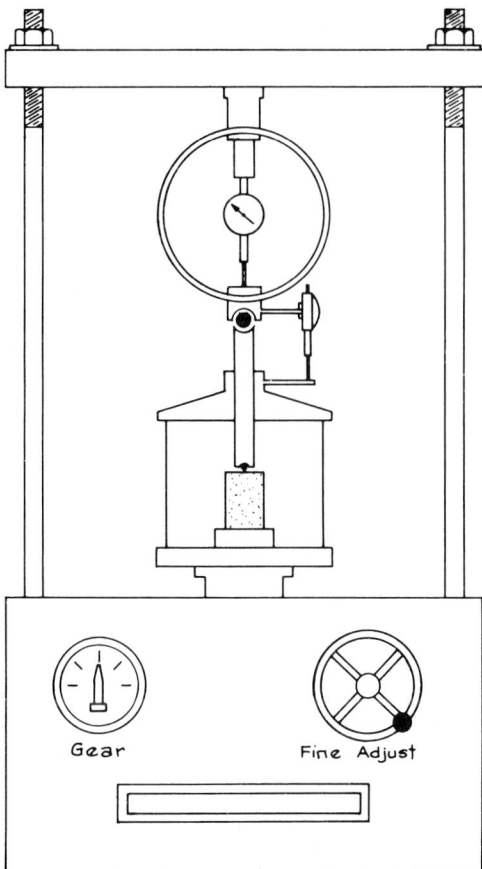

Fig. 4.19 Motorised loading frame for strain-controlled test

Fig. 4.20 Controlled-strain triaxial cell

The alternative type of loading system is that using hanger weights (as with the normal stress applied in the shear box), perhaps supplemented by a lever mechanism for large compressive loads. The load is incrementally applied at time intervals to suit the type of test being undertaken, i.e. relatively quickly in undrained triaxial tests and sufficiently slowly to allow dissipation of excess pore water pressures in drained triaxial tests.

A typical arrangement is shown in Fig. 4.21. This type of loading gives triaxial tests that are 'stress-controlled' in that the stress is being increased by specific increments at a chosen rate and this creates the subsequent strain. It is advantageous to use small, frequent load increments so that significant control is exercised. A stress–strain curve, plotted for the test as it proceeds, provides a measure of the sensitivity of control.

Drainage Facilities and Volume-change Observations
To allow observation of total volume change of the sample during the consolidation and/or shearing stage of a triaxial test, a drainage connection is fitted allowing continuous contact of the pore water within the sample to a burette or other volume-measuring device.

Often a saturated porous ceramic disc is placed between the top of the sample and the upper Perspex platen. This platen is grooved and drilled to connect to a rigid PVC tube, contained within the triaxial cell, and the tube connects the top of the Perspex platen to a drainage outlet in the cell base plate (Fig. 4.23).

From a Klinger valve in the cell base plate, a connection then leads to the base of a simple burette (measuring to 0·1 c.c.) attached by spring clips to one of the clamping rods on the outside of the triaxial cell. The clips allow the level of the burette to be adjusted continually so that the level of the water in the burette remains sensibly at the mid-height of the sample (the pressure datum). It is good practice to cover the surface of the water in the burette with a little olive oil so as to prevent evaporation, especially in long-term tests.

Back-pressure Systems
In some cases a back-pressure is connected to the drainage outlet, the function of which can be simply described as to equate levels of test pore-water pressure to field or *in situ* pore-water pressure. Two further distinct advantages are that a back-pressure (*a*) dissolves any residual air in the pipework connections and in the sample, thus ensuring full saturation, and (*b*) prevents negative gauge pressures and cavitation effects when testing soils that tend

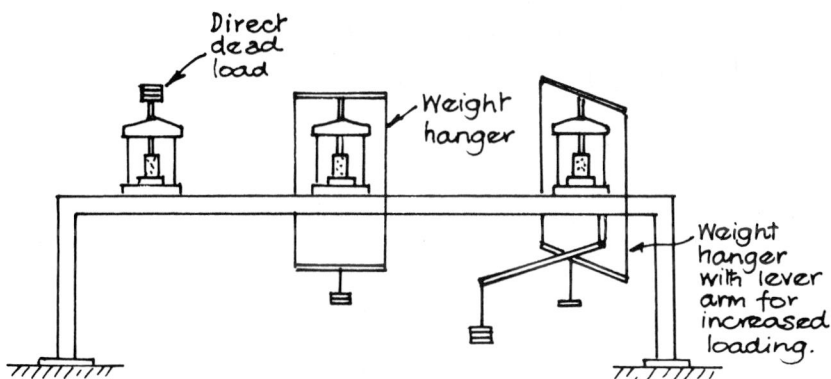

Fig. 4.21 Dead loading frame for stress-controlled test

to dilate (increase in volume) during shear, thus putting the pore water into tension. This applies particularly to certain overconsolidated soils.

A diagrammatic representation of draining against a back-pressure is shown in Fig. 4.24. The back-pressure can consist of a second Bishop constant-pressure mercury pot system or an elevated water-bottle with over-flow set to the *in situ* pore-water pressure. If, for example, a sample has been obtained from a site where the water table is at, say, 2 m below ground level

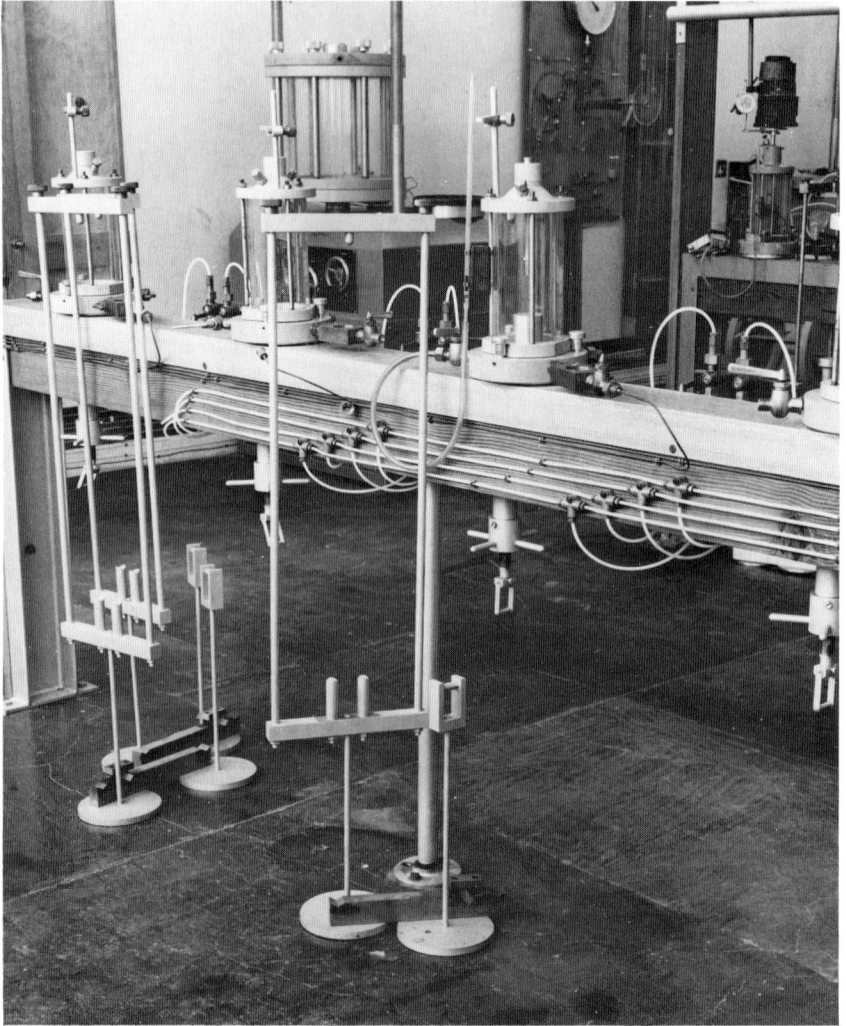

Fig. 4.22 Controlled-stress triaxial cell

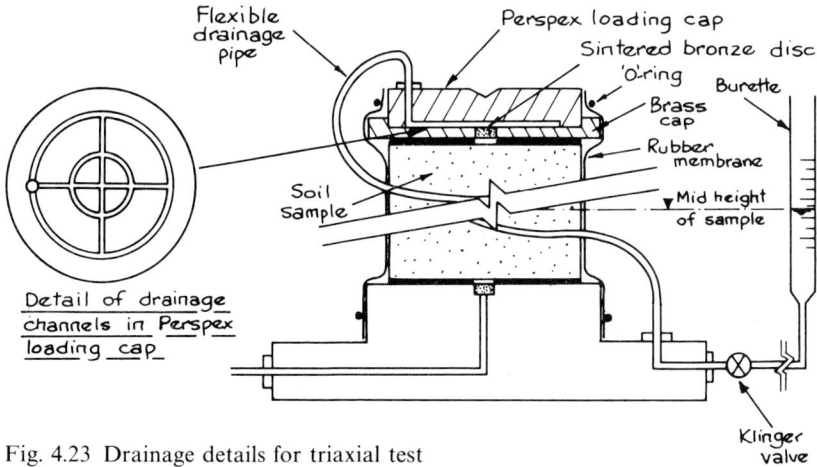

Fig. 4.23 Drainage details for triaxial test

and the sample depth below ground level is 5 m, then the existing pore-water pressure at the depth of sampling is $h \times \gamma_w = (5 - 2) \times \gamma_w$ where h = depth below the water table. Assuming that $\gamma_w = 10$ kN/m³, the pore-water pressure is $(5 - 2) \times 10 = 30$ kN/m². The back-pressure system is set to correspond to this pore-water pressure. Any pore pressures set up by loading the sample will be in excess of this value and resultant dissipation, after consolidation or drainage, will be complete when the pore-water pressure regains the 30 kN/m² value.

When testing with a back-pressure, volume change observations are undertaken by introducing a fluid that is immiscible with water into a burette that is capable of both withstanding the back-pressure without leakage and accommodating the total volume change of the sample. Several fluids are used for this purpose, kerosene being perhaps the most common.

Setting up the burette and back-pressure is accordingly rather more complex but once it is operational, little further difficulty is usually encountered.

Vertical Deformation
Usually vertical deformation of the sample is monitored by a dial gauge recording vertical movement of the loading piston (Fig. 4.25). In some cases lineal displacement transducers can also be installed to record vertical deformation. The transducers are electrically excited and can be installed to print out automatically the deformation of the sample at specified time intervals. This refinement can thus save attendant-time and is obviously of benefit in long-term tests running day and night.

4.5.2 Measurement of Pore-water Pressure

The fundamental reasons why care is needed in the measurement of pore-water pressure are the following:

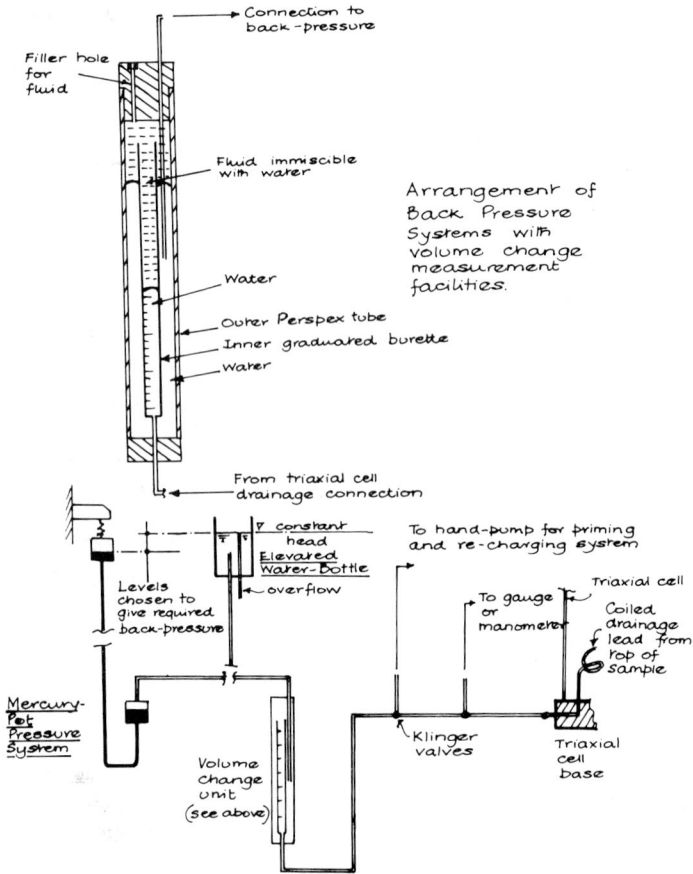

Fig. 4.24 Arrangement of back-pressure system with volume-change measurement
facility

1. There must be no volume change within the measuring system. This
   means that conditions of 'no flow' must exist in the connections from
   the base of the soil sample to the pore-water pressure measuring device.
   Consequently it is essential that all such connections are filled with de-
   aired water, regularly pressurised, flushed and tested for the presence of
   air before assembling and testing samples. If flow does occur due to
   volume changes, there will be a modification to the pressure observed
   and such observed values will not be accurate.
2. Unequal stress distributions and pore-water pressure distributions can
   develop throughout the height of triaxial samples because of end-
   restraint imposed by the loading platens in contact with the sample.
   Thus measured pore-water pressure at the base of a sample may not be

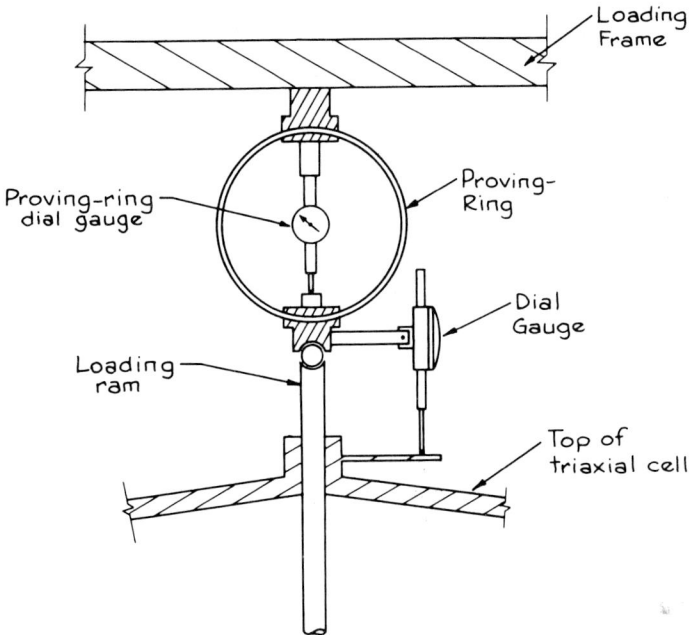

Fig. 4.25 Proving-ring and dial-gauge set-up on triaxial cell

the value of pore-water pressure at other points within the height of the sample.

Two common methods of measuring the pore-water pressures existing in triaxial samples under test are (i) use of a null indicator system and (ii) use of a pressure transducer.

Null Indicator System

Figure 4.26 shows diagrammatically the set-up of a null indicator system connected to the base of a triaxial cell.

The pore-pressure connection leaves the cell base at a Klinger valve and is then led in copper tube (usually 3 mm outside diameter by 1·5 mm internal diameter) to a brass connection fitted to the null indicator formed from 25 mm thick Perspex and mounted so that it can be rotated. The null indicator is essentially a U-tube that is partially filled with mercury. The pore-pressure lead then leaves the null indicator in copper tubing and is fed into four Klinger valves leading to a pressure gauge, a hand-screw pump and a mercury manometer.

Before testing, the whole circuitry is filled with de-aired water and constantly pressurised and flushed through the pore-pressure outlet in the triaxial cell base. While this is being done the Perspex null indicator is rotated on its mounting so that water is able to pass freely over the surface of the mercury.

The reason why copper tube is used is that it has zero volume-change characteristics (or very nearly so) so that pressures are transmitted without any modification. Klinger valves operate in a similar manner when opened and closed.

When pore-water pressure observations are required, the null indicator is rotated on its mounting and, by means of the hand-pump, mercury is very carefully pumped up the fine drilled hole to form a thread about 12 mm high. If the system is now opened to allow contact with the pore-water pressure from the sample, flow of pore water causes movement of the mercury thread. Careful adjustment of the water pressure on the other side of the mercury by means of the hand-pump can move the mercury back to its original position, which is marked. The pressure required to maintain the mercury at this marked position (i.e. in a condition of 'no flow') can be monitored on the pressure gauge or, for small pressures, on the mercury manometer.

The technique is soon acquired and the essence of successful operation is care at all stages. As the pore pressure within the sample rises or falls, constant adjustment by pump is required: the method of measurement is thus manual and requires constant technician attendance, which can be a handicap in some circumstances, whereas the transducer method of measurement has the advantage of automatic recording.

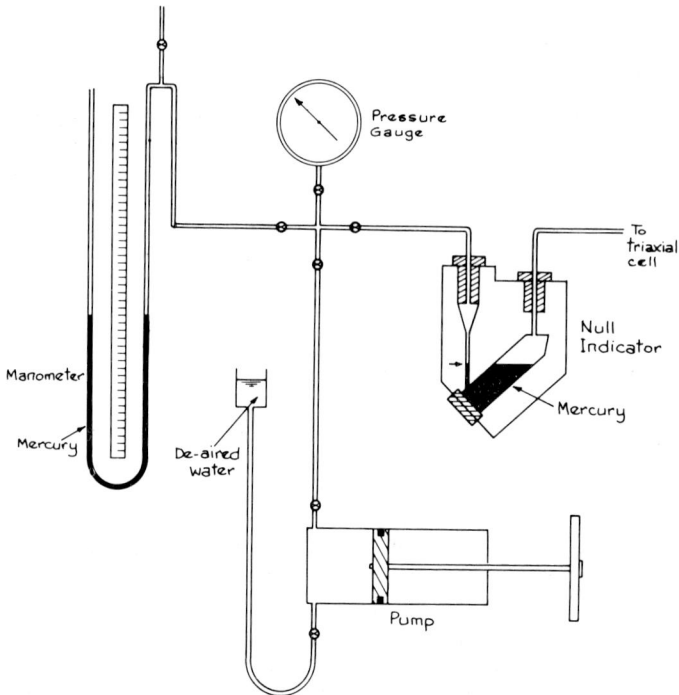

Fig. 4.26 Typical null indicator set-up

Pore-water Pressure Transducers

A variety of transducers is now available commercially. The transducer used with success for the measurement of pore-water pressures is the stitched diaphragm type.

The diaphragm is in contact with the pore water via a connection from the base of the sample to the transducer housing, which is preferably located close to the base of the sample. The transducer is supplied with an electric current and, as the diaphragm deforms under pressure, the voltage difference induced by such deformation is recorded on a digital voltmeter with visual display.

Thus by subjecting the transducer to a range of pressures and recording the voltage generated by each pressure, a calibration graph relating water pressure to voltage can be drawn.

Further refinements can be the use of digital voltmeters capable of receiving signals from several transducers, each assigned to a channel, the periodic scanning of these channels at predetermined times and the facility for printing out the results of such scanning operations together with a print-out of the real times of scanning. Display of the voltages is often in millivolts and scanning intervals can vary from, say 2 seconds up to 10 hours; thus one can have a continual print-out of the pressure–time relationship at predetermined intervals.

The increasing popularity of transducers is due to two major features:

Automation of the measuring process
Rapid response to rapidly changing pressures.

Figure 4.27 shows a typical transducer mounted to a triaxial cell, together with a digital voltmeter. The arrangement of a digital voltmeter serving several transducers is shown in Fig. 4.28. Once again, all associated pipework is made in 3 mm O.D. copper tube and all valves are Klinger valves so that negligible volume change is ensured. The whole circuitry is filled with de-aired water, pressurised and flushed by means of the screw control pump. A null indicator is also usually incorporated as a check for the presence of air.

Both of the systems described above – null indicators and transducers – generally give very acceptable pore pressure measurements and satisfy the condition of 'no flow' and consequent absence of pressure modification.

'Free' Ends Arrangement

To equalise the distribution of stresses and pressures throughout the depth of the cylindrical triaxial sample some effort should be made to reduce the end-restraint and frictional effects of the top and bottom platens that are in contact with the sample. Rowe and Barden[7] present some excellent ideas on the importance of 'free' ends in triaxial testing. 'Free' ends can be of distinct advantage in that they allow samples to be tested having height-to-diameter ratios of unity rather than the generally used ratio of 2:1, so that a greater number of test samples is produced from a given length of sampling. This is

Fig. 4.27 Triaxial cell with pore-water pressure transducer and digital voltmeter

especially important with larger diameter samples since it allows testing of 100 mm high × 100 mm diameter samples.

Usually the 'free' ends are as detailed in Fig. 4.29. The advantages of this arrangement can be summarised as follows:

1. Sample 'barrelling' – obvious indication of non-uniform stress distributions – is reduced and samples maintain their cylindrical shape.

2. More uniform results are obtained and samples with a height-to-diameter ratio of 1 are more stable, especially with 100 mm and 254 mm diameter samples.
3. Strain and pore-water pressure distributions are improved.

The technique of modifying existing platens is simple and, since the benefits of this system are so marked, there is every reason to use 'free' ends in all testing.

4.5.3 Setting Up a Triaxial Sample and Application of the Cell Pressure

The sample is extruded from the sampling tube, in which it has been stored, and trimmed to suit a split mould of the required sample size. The mould is usually manufactured from brass, is split along two diametrically opposed generators and held as a cylinder by a brass sleeve fitting over it. The split

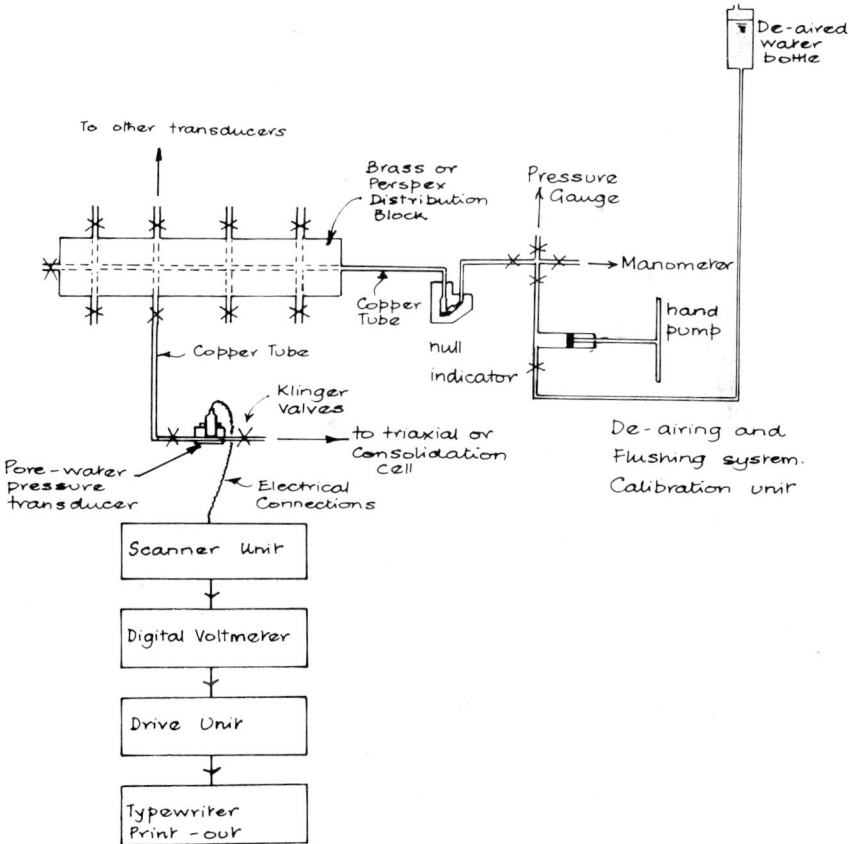

Fig. 4.28 Layout of digital voltmeter serving several transducers

mould containing the sample is then mounted on the base pedestal of the cell. A rubber sleeve or membrane of the appropriate size is stretched in a 'membrane stretcher', which is simply a tube with a pipe connection at its mid-height to enable a suction to be applied to the air gap between the membrane and the tube membrane stretcher. Both the split mould and the membrane stretcher are shown in Fig. 4.30.

The split mould is removed from the sample and the stretched membrane is placed around the sample and over the base pedestal. The top platen is then carefully placed into contact with the top of the sample, the suction on the membrane is released and totally encloses the sample and the sides of the upper and lower platens. 'O'-ring seals are then placed over the membrane by using a cylindrical stretcher and these rings are carefully rolled from the stretcher to the sides of the lower and upper loading platens, care being taken not to disturb the sample. The rings then securely clamp the membrane and prevent any seepage of water between the membrane and the sides of the platens. With larger samples it may be necessary to use two 'O'-rings at both the top and the bottom of sample and, in addition, to clamp the 'O'-rings by using a hose or jubilee clip.

It is essential that any connections to drainage facilities or pore-water pressure measuring devices are flushed through before assembling the sample and that the associated tube connections are full of water. Any porous discs used must be saturated by boiling in water.

The Perspex cell is then assembled over the sample and securely tightened

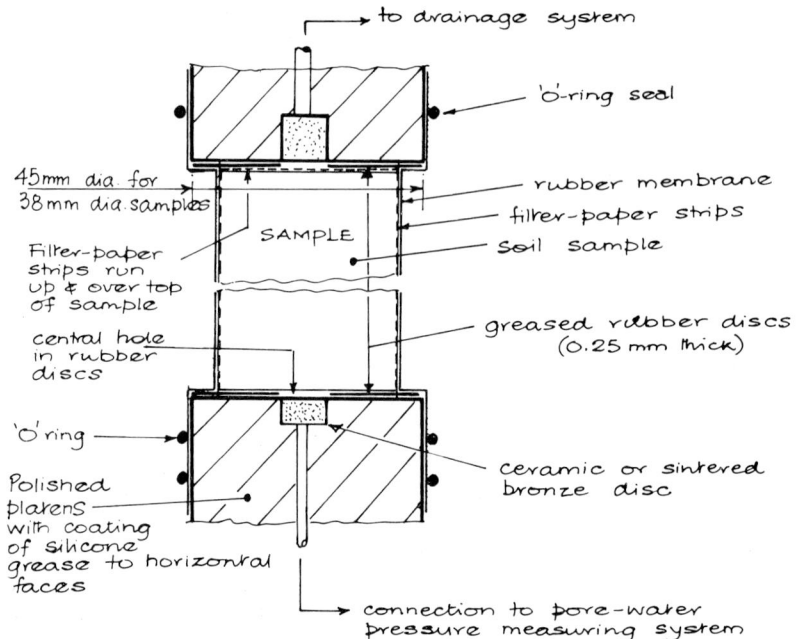

Fig. 4.29 'Free' ends arrangement

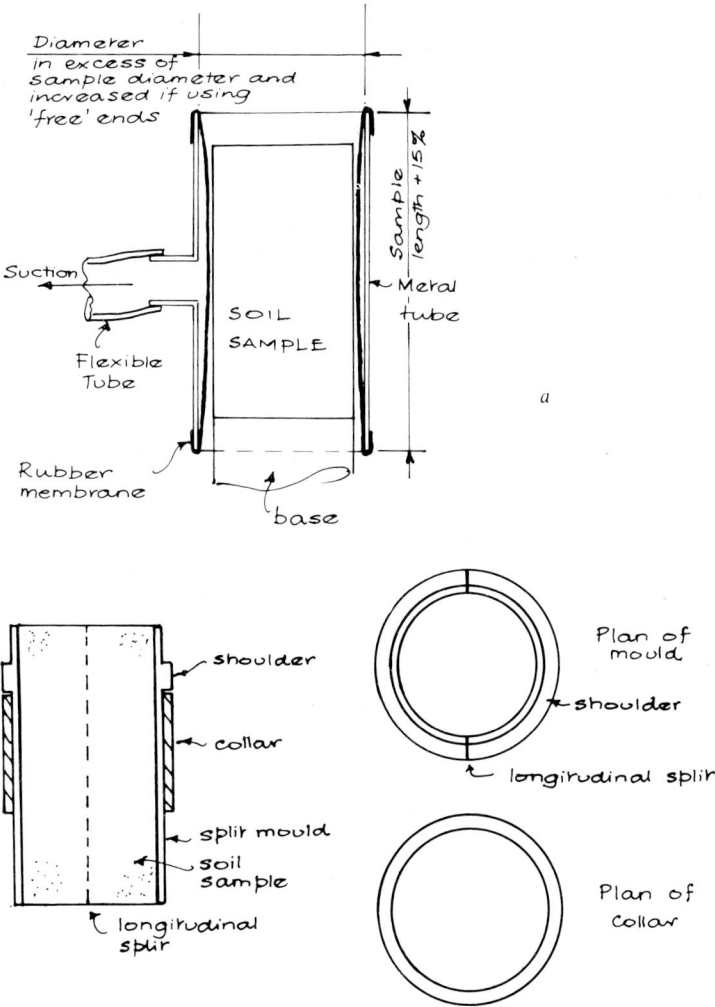

Fig. 4.30(a) Membrane stretcher
        (b) Split mould (brass)

to the base plate of the cell. The piston or ram is carefully lowered to rest on the ball seating on top of the sample, clamped in such a position that it just touches the ball seating. Extreme care is needed at this stage to ensure that no vertical load is applied to the sample.

The triaxial cell is filled with water from a header bottle, the air release or bleed valve at the top of the cell being kept open to allow filling. When the cell is full this valve is closed. The cell is then pressurised by a mercury-pot system, or from an air–water cylinder, preset to the required equal all-round cell pressure or confining pressure.

Depending upon the type of testing being undertaken, drainage can now be allowed into the burette via the drainage outlet at the top of the sample. This is the consolidation stage, and progress is recorded by monitoring both the pore pressure and burette readings as drainage proceeds. When testing with zero back-pressure it is important to adjust the level of the oil-covered water in the burette constantly, by lowering the burette during consolidation, so that the water level remains at the mid-height of the sample. If a back-pressure system is used, this is maintained at a constant pressure level by an appropriate means. Consolidation is complete when the volume change has ceased and the pore-water pressure, recorded on the measuring device, corresponds to the back-pressure (at a predetermined level or at zero).

The time required to reach a consolidated state depends upon the type of soil being tested and upon the size of the sample. The consolidation time is proportional to the square of the length of drainage path. Thus, with purely one-way vertical drainage up the sample to the saturated porous disc incorporated into the top loading platen, the comparative times to reach a given degree of consolidation for a 100 mm dia $\times$ 100 mm high sample and a 38 mm dia $\times$ 76 mm high sample of the same soil are

$$\frac{t_{100}}{t_{38}} = \frac{(H_{100})^2}{(H_{38})^2} = \frac{100^2}{76^2} = 1.73$$

i.e. in theory, consolidation of a 100×100 mm sample takes 1·73 times longer than that of a 38×76 mm sample. If the 100 mm diameter sample were 200 mm high, so that the height to diameter ratio was 2, the comparative ratio of consolidation times would be

$$\frac{t_{100}}{t_{38}} = \left(\frac{200}{76}\right)^2 = 6.9$$

This illustrates a further advantage of the use of 'free' ends to allow the testing of samples with height-to-diameter ratios of unity, giving substantially quicker consolidation rates.

An additional means of accelerating consolidation rate is to attach filter-paper strips to the cylindrical surface of the sample (Fig. 4.31). This technique greatly increases consolidation rates since it allows drainage of pore water both radially and vertically within the sample. The radial consolidation to the filter-paper strips is now along a drainage path length equal to the radius of the sample and, in addition, some vertical drainage occurs. However, a correction may have to be made for the restraining effect of the filter-paper drains during the shearing stage of the sample. For this correction, together with full details of specialist triaxial techniques, the reader is referred to Bishop and Henkel.[8]

Choice of Cell Pressure
The choice of magnitude of cell pressure is usually such that the effective pressure of the sample after consolidation is the same as the vertical effective

pressure existing on site at the depth of sampling. For example, suppose sampling is undertaken at a site where, at the proposed depth of sampling, total vertical pressure = 100 kN/m² and pore-water pressure = 30 kN/m², so that

$$\text{Effective vertical pressure} = 100 - 30 = 70 \text{ kN/m}^2$$

The effective laterial pressure at the site, at sample depth, may not necessarily be equal to 70 kN/m² since it depends upon K_0, the coefficient of earth pressure at rest. Suppose that $K_0 = 0.6$, therefore

$$\text{Effective lateral pressure} = K_0 \times 70 = 0.6 \times 70$$
$$= 42 \text{ kN/m}^2$$

Thus the state of the sample in terms of effective and total pressures is as shown in Fig. 4.32.

The sample is then taken, the ends of the sample tube are sealed with wax and it is taken to the testing laboratory. Assuming that no moisture enters or leaves the sample at any stage (i.e. its moisture content remains constant) and that sample disturbance has been minimal so that there is no distortion and consequently the sample retains the same volume, then, when it is extruded in the laboratory and immediately enclosed in a rubber sheath and end-platens, externally applied stresses on the sample will be zero. The state of the sample is then as shown in Fig. 4.33, and to maintain average effective pressure levels the pore-water pressure u thus needs to be negative (suction).

The function of the consolidation part of the test is to attempt to simulate effective pressures on site, therefore the cell pressure should be set accordingly. This can be done in more than one way. In the sample under consideration, effective pressure required = vertical effective pressure at field scale = 70 kN/m². Therefore if cell pressure (total pressure) = 70 kN/m², back-pressure (pore-water pressure) = 0 kN/m², and field values of effective pressure are achieved by draining to a burette set with the water level at the mid-height of the sample (i.e. zero back-pressure).

More realistic values could, however, be achieved in this particular case by setting cell pressure = 100kN/m² and back pressure = 30kN/m² (achieved

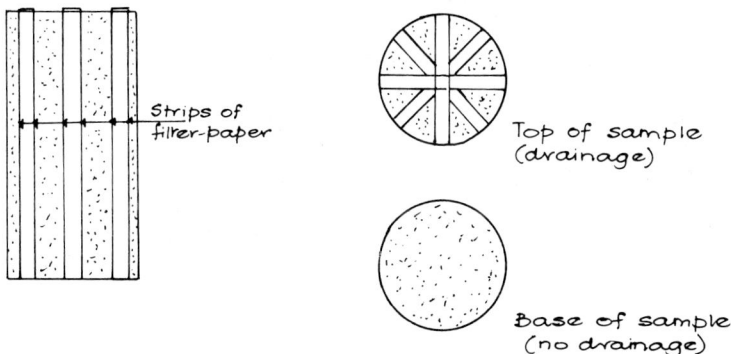

Strips of filter-paper

Top of sample (drainage)

Base of sample (no drainage)

Fig. 4.31 Filter-paper strips used to accelerate consolidation rate

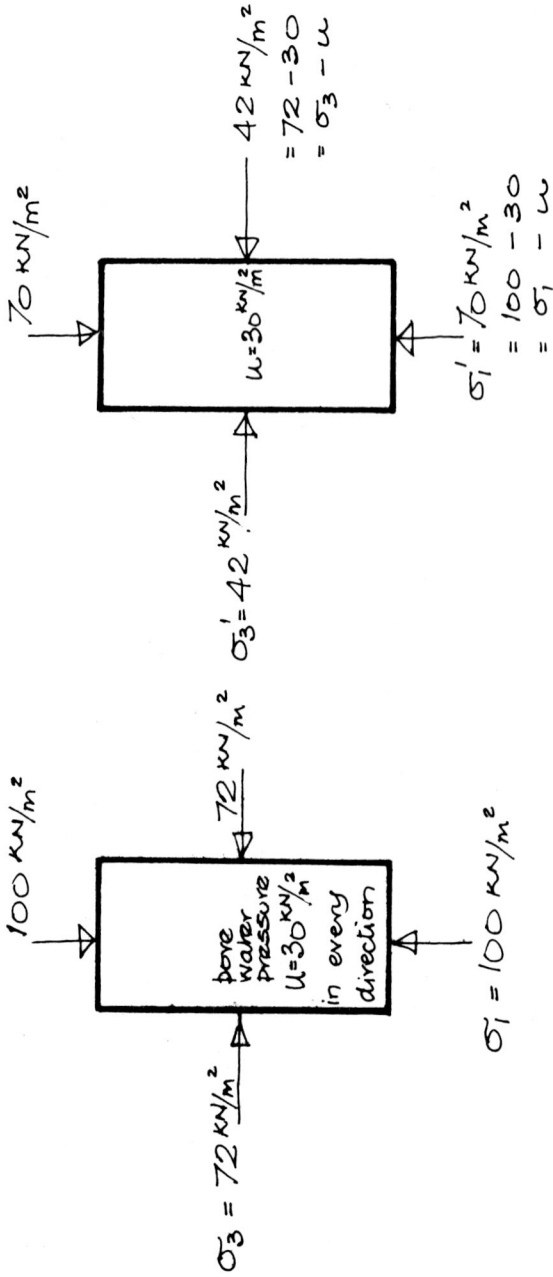

(a) Total pressures

$\sigma_3 = 72$ KN/m²

$\sigma_1 = 100$ KN/m²

100 KN/m²

72 KN/m²

Pore water pressure $u = 30$ KN/m² in every direction

(b) Effective pressures

70 KN/m²

42 KN/m² $= 72 - 30 = \sigma_3 - u$

$\sigma_3' = 42$ KN/m²

$u = 30$ KN/m²

$\sigma_1' = 70$ KN/m² $= 100 - 30 = \sigma_1 - u$

Fig. 4.32(a) Total pressures

by means of a mercury pot system or similar), so that the resultant effect-ive pressure $= 70 \text{kN/m}^2$.

With these values of pressure not only is a similar state achieved but also similar levels of the appropriate pressures are operative. The added advan-tages provided by such a back-pressure are that any residual air is dissolved into solution, ensuring full saturation of the sample, and, if during shearing there is a tendency for the soil to dilate, then the risk of negative gauge pressures and subsequent cavitation is reduced.

The choice of correct and most appropriate cell pressure for testing is a necessary preliminary to any series of triaxial testing; also, to produce a set of three Mohr's circles and the resultant failure envelope, it is necessary to choose an appropriate range of cell pressures. It is important that this range correlates to expected or actual field values.

4.5.4 Application of Shearing Load or Deviatoric Load

After the sample is set up in the triaxial cell and, if required, consolidated to a required effective stress, the next stage is to create a principal stress differ-ence on the sample by increasing the vertical stress acting on its top and bottom circular faces. This is achieved by increasing the load on the piston or ram at the top of the sample.

A very important aspect of this part of the test is the *rate* at which this load is increased. The three main types of test usually involve various load-ing rates and have been described in terms of time as 'quick' or 'slow' tests. Such descriptions are misleading since it is important to realise that times are relative to the magnitude of the problem under consideration. For example, a drained triaxial ('slow') test can be performed satisfactorily on some soils in less than one day while other soils may need one month or more for adequate performance. Similarly, with a clay that is of low per-meability and has consolidation characteristics that indicate dissipation of pore-water pressures over perhaps 20 or 30 years or more, the application of a load to the soil over 2 or 3 months at the field scale is effectively a 'quick' process, there being very little dissipation of excess pore-water pressure during such a loading period. Thus strength analysis for stability calcula-tions at 'the end of construction' should, in this case, be based on undrained strength and consequently undrained triaxial tests are appropriate and relevant.

Generally, the rate of strain used for undrained and consolidated-undrained triaxial tests is 2% per minute. For drained tests the rate of strain is chosen so that there is no build-up of excess pore-water pressure during shearing. Therefore the rate chosen depends very much on the permeability of the soil and the rate at which excess pore-water pressure can be dis-sipated.

An appropriate rate of strain having been chosen, then, if automatic load-ing frames are used, the corresponding feed rates are chosen, usually by selecting a particular gear and fixing the corresponding gear wheels. The loading frame then moves the whole triaxial cell upwards at the chosen rate

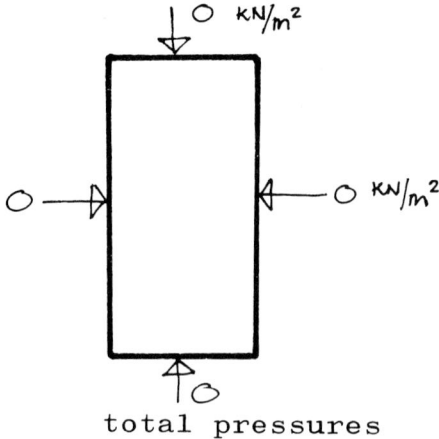

0 kN/m^2

0 kN/m^2

0

Fig. 4.33

total pressures

so that the loading ram, resting on the top of the sample, is pushed against a proving-ring against a beam reaction. This increases the load on the sample, as strain is induced. It is obviously necessary to choose a proving-ring that is appropriate to the anticipated load range. In addition the proving-ring needs to be well-maintained and to have been recently calibrated so that observed readings of the proving-ring dial gauge correlate accurately with loading.

It is also important that the loading ram which passes through the top of the triaxial cell should be as friction-free as is compatible with providing a watertight seal (so that the confining water pressure is maintained within the cell). The ram should be very carefully polished, lightly greased and able to move freely.

The compression of the sample is continued at a steady strain rate until failure occurs. Usually failure is signified by a fall-off in applied vertical load with increase in compressive strain; if there is no apparent reduction in load, the failure load is taken as that load corresponding to 20% strain.

Dependent upon the type of test being undertaken, observations of vertical deformation, proving-ring dial gauge, pore-water pressure, volume-change burette and checks on the lateral confining pressure are recorded throughout the testing. It is also good practice to plot a graph of principal stress difference versus strain during testing. This is particularly useful in drained tests.

It must be noted that principal stress difference is the stress applied via the loading ram (Fig. 4.34). $\sigma_1 =$ major principal stress and $\sigma_3 =$ minor principal stress; thus $\sigma_1 - \sigma_3 =$ principal stress difference. But σ_1 is made up of two components: (i) the cell pressure applied equally all round the sample, plus (ii) the axial stress induced by the loading frame; that is

$$\sigma_1 = \sigma_3 + \frac{P}{A}$$

(cell   (axial
pressure)   stress)

where P = applied vertical load recorded by proving-ring, and A = cross-sectional area of the sample which varies with vertical deformation. Thus

$$\sigma_1 - \sigma_3 = \frac{P}{A}$$

and consequently the evaluation of $(\sigma_1 - \sigma_3)$ involves a series of computations to evaluate A, the cross-sectional area of the sample.

The average cross-sectional area A of the sample is given, at any stage of testing, by the expression

$$A = A_0 \frac{1 - (\Delta V / V_0)}{1 - (\Delta l / l_0)}$$

where ΔV = change in volume measured by burette if drainage is allowed
  V_0 = initial volume (before shearing but after consolidation)
  Δl = change in axial length measured by dial gauge monitoring loading-ram movement
  l_0 = original length (before shearing but after consolidation)
  A_0 = original cross-sectional area of the sample (before shearing but after consolidation)
  $\Delta V / V_0$ = volumetric strain
  $\Delta l / l_0$ = axial strain.

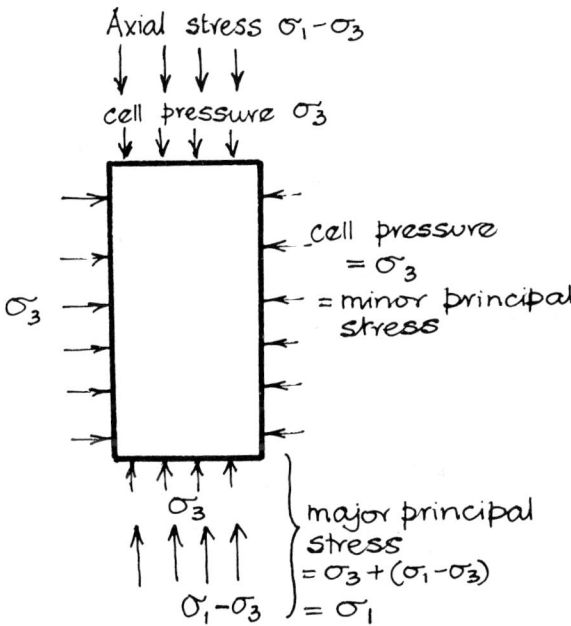

Fig. 4.34 Stress applied via loading ram

Typical results are given in Datasheet No. 12, including full calculations together with stress–strain plots.

One important factor that may lead to erroneous results is the amount of friction between the loading ram and the top of the triaxial cell. An obvious advantage is to measure the load applied directly to the top of the sample within the triaxial cell. This involves using load cells incorporated into the loading ram; because they are in contact with water, such load cells need to be unaffected by immersion in water. Usually such load cells utilise strain gauges which can be monitored by means of a digital voltmeter, connected via a hollow loading ram.

Other requirements are that the load cells must be calibrated, be of sufficient sensitivity and linearity, be of an appropriate range and be stable under water. With smaller samples the physical size of the load cell is limited and the triaxial cell must be capable of accepting the modified loading rams. Obviously to provide a comprehensive range of load cells giving a wide range of load capacities is expensive but their use is justified where results of high quality are required.

It is also of interest to note that BS 1377: 1975 specifies that 'a correction to allow for the restraining effect of the rubber membrane shall be made' and details the correction in Note II to Test 21 of the Standard. Allowance for the effect of the membrane results in a deduction from the measured maximum principal stress difference. In addition, BS 1377:1975 gives details of the apparatus required to determine the extension modulus of the rubber membrane, based upon a method proposed by Henkel and Gilbert.[9]

When drained tests are being undertaken, it is important that the pore-water pressure within the sample is kept under continual observation, so as to ensure that no excess pore-water pressure is developed during the shearing process. The rate of loading is chosen to ensure that this condition is satisfied. Another feature of the drained test is that the soil is continually decreasing in moisture content due to the expulsion of pore water, of which the quantity is recorded by the burette system. Consequently, in the calculation of vertical stresses applied to the sample, account must be taken of this continual volume change when computing the appropriate cross-sectional area of the sample at any stage of testing. An example of such calculations is given in Datasheet No. 13 (page 114).

In all types of conventional triaxial test, the vertical load is increased until there is a clear and marked fall-off in the value of the applied deviator stress (usually signified by continual axial strain with no increase in deviator stress) or until the axial strain has reached 20%, which usually occurs with a plastic-type failure. The value of the deviator stress that is used in drawing the appropriate Mohr's circles and subsequent evaluation of the shear strength parameters is thus either the maximum deviator stress recorded or the deviator stress recorded at 20% axial strain.

4.5.5 Summary

1. In spite of some inherent weakness, the triaxial test is still used on a

Undrained Triaxial Test.

Sample Number	Mass of Natural Sample (g)	Mass of Dry Sample (g)	Moisture Content (%)	Bulk Density (mg/m³)	Cell Pressure σ_3 (kN/m²)	Sample Size (mm)
1	159.9	137.0	16.7	2.26	140	63.5×38
2	157.1	134.5	16.8	2.22	280	63.5×38
3	159.5	136.2	17.1	2.25	420	63.5×38

Strain Dial Gauge	Divisions Proving Ring Dial Gauge	Divisions Proving Ring Dial Gauge	Divisions Proving Ring Dial Gauge	Strain %
·636	22	15	35	1
1·270	32	28·5	43	2
1·905	38·5	34·25	49	3
2·540	43·5	38·75	53·8	4
3·175	48	43	58	5
3·810	51·5	46·75	61·7	6
4·445	55	50·2	64·6	7
5·080	57·75	53·5	67·2	8
5·715	60·0	56·8	70	9
6·35	63·25	59·6	72	10
6·985	64·5	62·5	73·8	11
7·620	66·25	65	75·5	12
8·255	68	67·3	77	13
8·890	70	69·7	78·5	14
9·525	71	71·9	79·8	15
10·160	72·5	73·7	81	16
10·795	74	75·6	82·1	17
11·430	74·6	77·2	83·1	18
12·065	75·3	78·8	84	19
* 12·70	76	80	84·7	20
13·335	76·7	81·2	85·5	21
13·970	77·1	82·6	86·1	22
14·605	77·5	83·5	87	23
15·240	77·8	84·1	87·2	24
15·875	78	85·0	88	25
16·510	78·1	85·5	88·5	26
17·145	78·3	85·9	89	27
mm.	$\sigma_3 = 140$ Test 1	$\sigma_3 = 280$ Test 2	$\sigma_3 = 420$ Test 3	

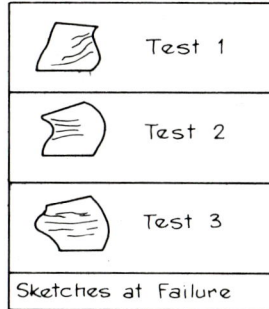

Test 1

Test 2

Test 3

Sketches at Failure

Proving Ring Constant from Calibration Chart
= 4.45 N/Division.

Failure Loads.

Test 1 = $76 \times \dfrac{4.45}{1000} = \cdot3382$ kN

Test 2 = $80 \times \dfrac{4.45}{1000} = \cdot3560$ kN

Test 3 = $84.7 \times \dfrac{4.45}{1000} = \cdot3769$ kN

Failure Stresses (Deviator = $\sigma_1 - \sigma_3$)

Test 1 = $\dfrac{\cdot3382}{\cdot00142} = 238.2$ kN/m²

Test 2 = $\dfrac{\cdot3560}{\cdot00142} = 250.7$ kN/m²

Test 3 = $\dfrac{\cdot3769}{\cdot00142} = 265.4$ kN/m²

* 20% strain taken as failure condition for plastic failure.
   C.S. Area at failure, $A = \dfrac{A_o}{1-\varepsilon}$ where A_o = initial C.S.A.
   $= \dfrac{\pi}{4} \times \left(\dfrac{38}{1000}\right)^2 = \cdot00113$ m²

   ε = strain = 0.2

$$A = \frac{\cdot00113}{1-0.2} = \frac{\cdot00113}{0.8} = \underline{\cdot00142 \text{ m}^2}$$

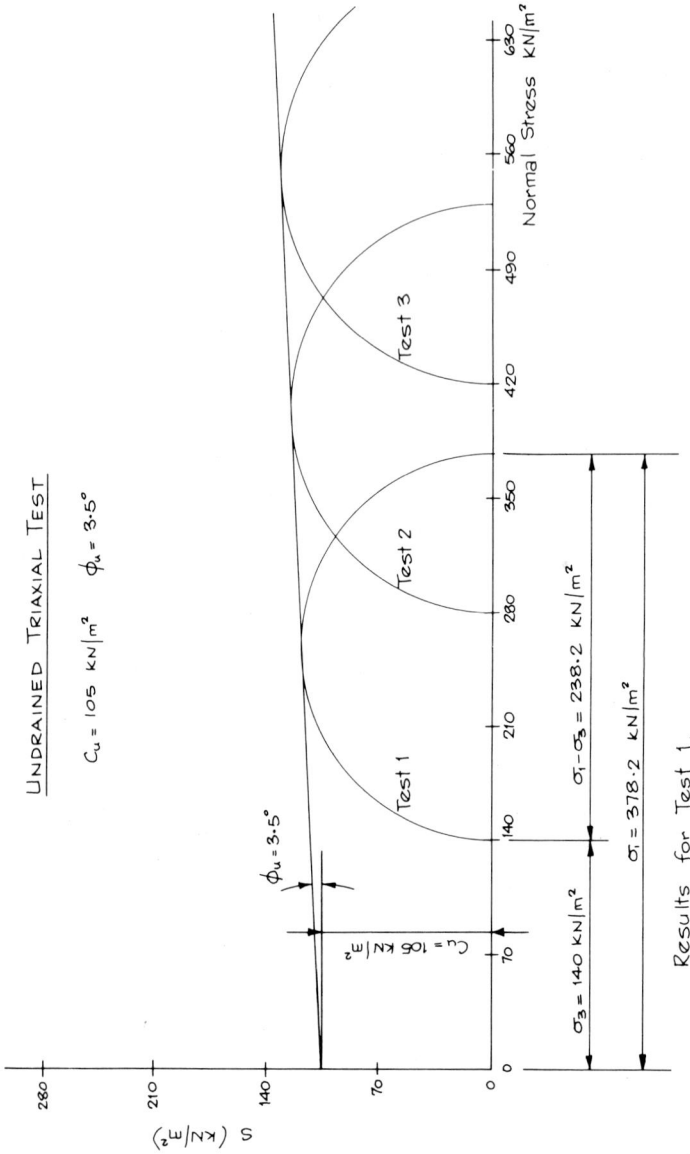

UNDRAINED TRIAXIAL TEST

$C_u = 105$ KN/m² $\quad \phi_u = 3.5°$

$\phi_u = 3.5°$

$C_u = 105$ KN/m²

Test 1

Test 2

Test 3

Normal Stress KN/m²

$\sigma_1 - \sigma_3 = 238.2$ KN/m²

$\sigma_1 = 378.2$ KN/m²

$\sigma_3 = 140$ KN/m²

Results for Test 1.

S (KN/m²)

Consolidated Undrained Triaxial Tests with Pore Water Pressure Measurements (using frictionless ends.)

Physical Properties.

SAMPLE	A1	A2	A3
Sample Size	63·5mm x 38mm diameter		
Mass - natural sample (g)	161	163·4	164·3
Mass - dry sample (g)	146·4	145·0	147·3
Moisture content (%)	10%	12·7%	11·6%
Bulk Density (Mg/m³)	2·220	2·308	2·271

Consolidation Data. - stage 1

SAMPLE	A1	A2	A3
Cell pressure (kN/m²)	140	280	420
Digital Voltmeter reading (mV)	5·11	12·92	20·6
Pore Water Pressure (kN/m²)	122·5	213·5	350
$B = \dfrac{\text{Pore Water Pressure}}{\text{Cell Pressure}}$	0·875	0·765	0·835

Area of sample $(A_o) = \dfrac{\pi}{4} \times 38^2 \times \dfrac{1}{10^6} = \cdot 0011398 \ m^2$

Initial length of sample $(\ell_o) = 63\cdot5\,mm$

Controlled Stress Type Tests.

* Note

To calculate deviator stress $(\sigma_1 - \sigma_3)$ for example, take results for sample A2 overleaf.

when   Load $= (39\cdot5\,kg) \times 9\cdot81 = 387\cdot5\,N = \cdot3875\,kN$

Force $= (Mass) \times (Acceleration)$

and corresponding cross-sectional area of sample $= \cdot0011778\ m^2$

Thus deviator stress $= (\sigma_1 - \sigma_3) = \dfrac{\cdot3875}{\cdot0011778} = 329\ kN/m^2$

and $\sigma_1 = (\sigma_1 - \sigma_3) + \sigma_3 = 329 + 280 = 609\ kN/m^2$

$\sigma_3{}' = \sigma_3 - u = 280 - 63 = 217\ kN/m^2$

$\sigma_1{}' = \sigma_1 - u = 609 - 63 = 546\ kN/m^2$

and $\dfrac{\sigma_1{}'}{\sigma_3{}'} = \dfrac{546}{217} = 2\cdot51$

Consolidated Undrained Triaxial Test with Pore Water Pressure Measurements.

SAMPLE A1 — $\sigma_3 = 140$ kN/m²

Net Load (kg)	$\Delta\ell$ Vertical Deformation mm x·01	ε Strain $=\frac{\Delta\ell}{\ell_0}$	$1-\varepsilon$	Area $=\frac{A_0}{1-\varepsilon}$ (m²)	$\sigma_1-\sigma_3$ (kN/m²)	σ_1 (kN/m²)	D.V.M. Voltmeter Reading (mV)	u Pore Water Pressure (kN/m²)	σ_3' (kN/m²)	σ_1' (kN/m²)	$\frac{\sigma_1'}{\sigma_3'}$	$A=\frac{u}{B(\sigma_1-\sigma_3)}$
0	0	0	1.00	·0011398	0	140	−1·88	0	140	140	1	0
22·44	160	·0252	·9748	·0011692	188·3	328·3	+0·10	45·5	94·5	282·8	3	0·276
27·15	205	·0323	·9677	·0011779	226·1	366·1	−0·07	30·1	109·9	336·0	3·06	0·152
31·70	272	·0430	·9570	·0011910	261·1	401·1	−0·41	24·5	115·5	376·6	3·26	0·107
36·30	347	·0546	·9454	·0012056	295·4	435·4	−0·79	17·5	122·5	417·9	3·41	0·0766
40·79	450	·0708	·9292	·0012267	326·2	466·2	−1·80	0	140·0	466·2	3·33	0·0
46·46	817	·1288	·8712	·0013083	340·9	480·9	−2·43	−10·5	150·5	491·4	3·265	−0·035
47·84	1383	·2180	·7820	·0014576	322·0	462·0	−2·71	−14·0	154·0	476·0	3·10	−0·0496
50·11	1920	·3020	·6980	·0016330	301·0	441·0	−2·88	−19·25	159·25	460·25	2·99	−0·073

SAMPLE A2 — $\sigma_3 = 280$ kN/m²

Net Load (kg)	$\Delta\ell$ Vertical Deformation mm x·01	ε Strain $=\frac{\Delta\ell}{\ell_0}$	$1-\varepsilon$	Area $=\frac{A_0}{1-\varepsilon}$ (m²)	$\sigma_1-\sigma_3$ (kN/m²)	σ_1 (kN/m²)	D.V.M. Voltmeter Reading (mV)	u Pore Water Pressure (kN/m²)	σ_3' (kN/m²)	σ_1' (kN/m²)	$\frac{\sigma_1'}{\sigma_3'}$	$A=\frac{u}{B(\sigma_1-\sigma_3)}$
0	0	0	1.00	·0011398	0	280	−0·12	0	280	280	1	0
21·16	76	·0120	·9880	·0011536	179·9	459·9	+2·79	42·0	238·0	417·9	1·76	0·305
30·54	126	·0199	·9801	·0011629	257·6	535·5	+3·37	52·5	227·5	463·0	2·12	0·268
39·50	205	·0323	·9677	·0011778	329·0	609·0	+4·07	63·0	217·0	546·0	2·52	0·251
49·26	322	·0507	·9493	·0012007	402·5	682·5	+4·23	66·5	213·5	616·0	2·88	0·216
58·03	475	·0750	·9250	·0012322	462·0	742·0	+3·05	45·5	241·5	696·5	2·88	0·129
67·00	801	·1260	·8740	·0013041	504·0	784·0	+1·86	24·5	256·5	759·5	2·97	
69·49	1070	·1690	·8310	·0013716	497·0	777·0	+2·73	38·5	269·5	738·5	3·06	0·065
71·46	1220	·1920	·8080	·0014106	497·0	777·0	+2·57	36·3	244·0	740·7	3·06	0·065

SAMPLE A3 — $\sigma_3 = 420$ kN/m²

Net Load (kg)	$\Delta\ell$ Vertical Deformation mm x·01	ε Strain $=\frac{\Delta\ell}{\ell_0}$	$1-\varepsilon$	Area $=\frac{A_0}{1-\varepsilon}$ (m²)	$\sigma_1-\sigma_3$ (kN/m²)	σ_1 (kN/m²)	D.V.M. Voltmeter Reading (mV)	u Pore Water Pressure (kN/m²)	σ_3' (kN/m²)	σ_1' (kN/m²)	$\frac{\sigma_1'}{\sigma_3'}$	$A=\frac{u}{B(\sigma_1-\sigma_3)}$
0	0	0	1·0	·0011398	0	420	−1·23	0	420	420	1	0
19·32	42	·0066	·9934	·0011474	165·2	585·2	+0·61	28·0	392·0	567·2	1·42	0·203
37·66	83	·0131	·9870	·0011548	319·9	739·9	+2·36	56·0	364·0	683·9	1·89	0·210
69·83	152	·0240	·9760	·0011678	586·6	1006·6	+4·14	84·0	336·0	922·6	2·75	0·172
74·38	288	·0454	·9546	·0011940	611·1	1031·1	+4·68	94·5	325·5	936·6	2·88	0·185
92·84	515	·0810	·9190	·0012403	734·3	1154·3	+2·48	59·5	360·5	1094·8	3·04	0·097
106·43	1475	·2320	·7680	·0014841	703·5	1123·5	—	—	—	—	—	—
108·80	1882	·2960	·7040	·0016190	659·4	1079·4	+1·37	42·0	378·0	1037·4	2·75	0·076
111·24	2170	·3420	·6580	·0017522	630·0	1050·0	+1·19	38·5	381·5	1011·5	2·65	0·0735

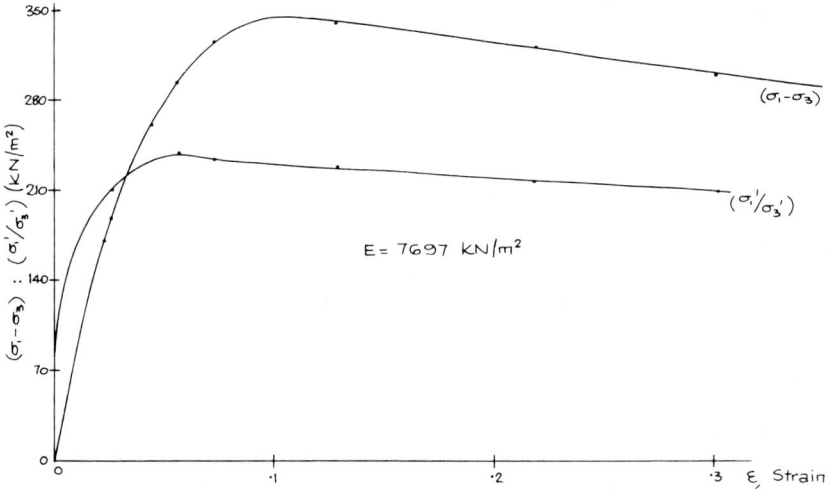

E = 7697 kN/m²

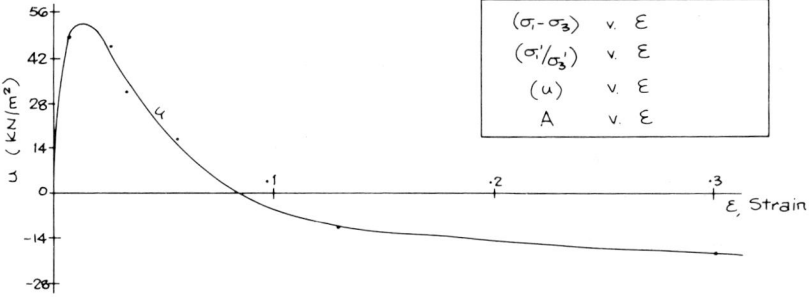

SAMPLE A1
Cell Pressure = 140 kN/m²

$(\sigma_1 - \sigma_3)$	v.	ε
(σ_1'/σ_3')	v.	ε
(u)	v.	ε
A	v.	ε

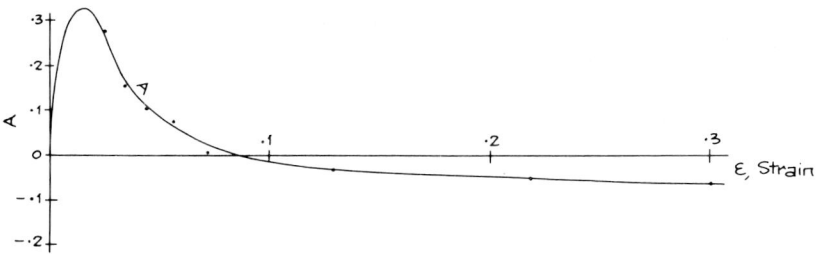

Datasheet Nº 12

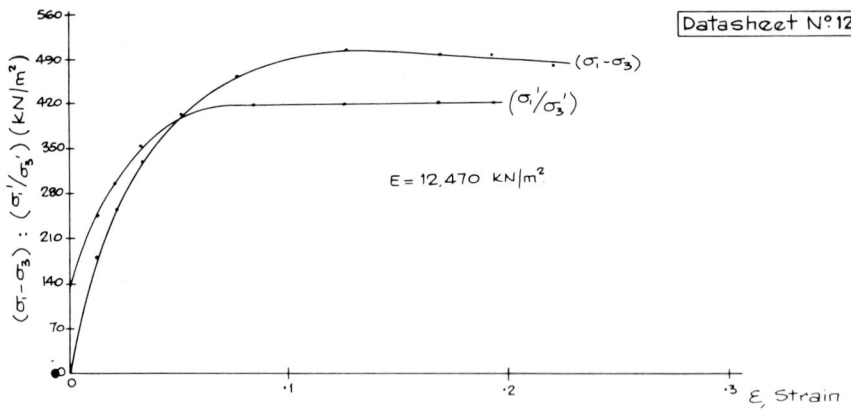

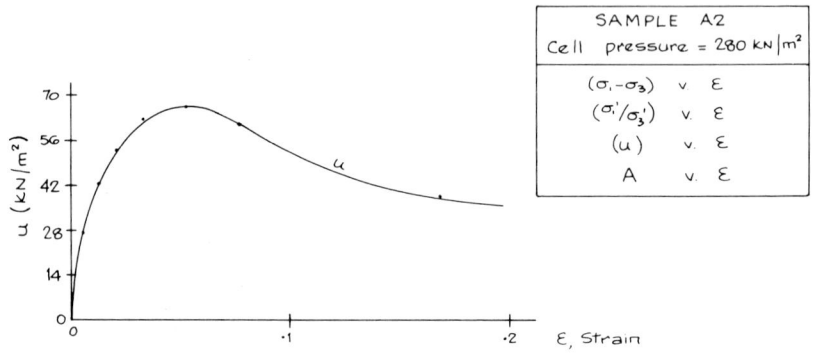

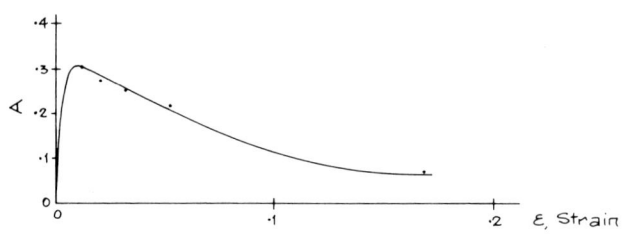

$E = 12,470 \ kN/m^2$

SAMPLE A2

Cell pressure $= 280 \ kN/m^2$

$(\sigma_1 - \sigma_3)$	v.	ε
(σ_1'/σ_3')	v.	ε
(u)	v.	ε
A	v.	ε

Datasheet N° 12

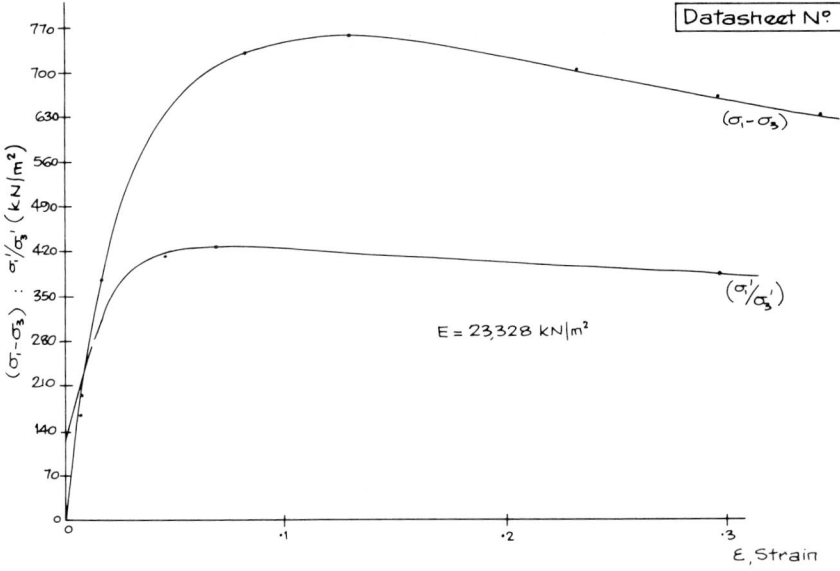

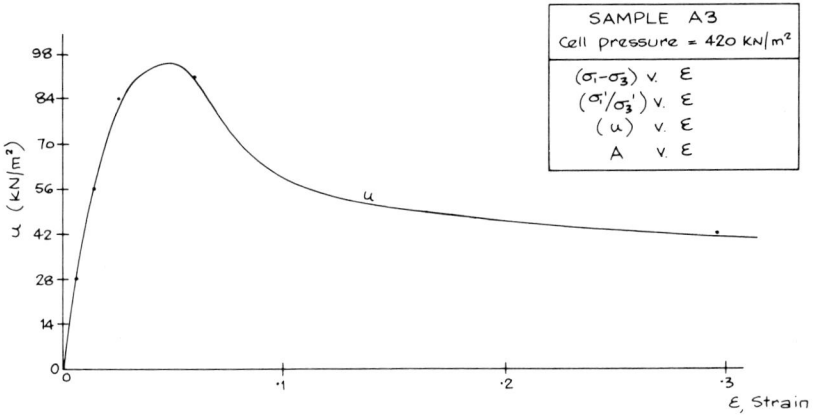

$E = 23,328 \ kN/m^2$

SAMPLE   A3
Cell pressure = 420 kN/m²

$(\sigma_1-\sigma_3)$ v. ε
(σ_1'/σ_3') v. ε
(u) v. ε
A v. ε

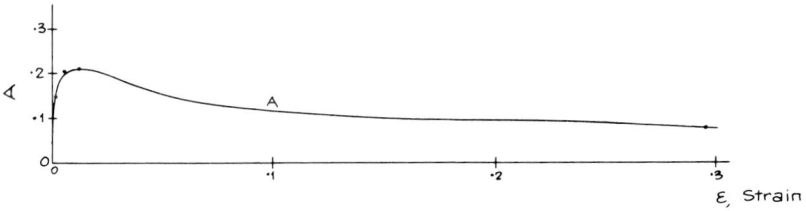

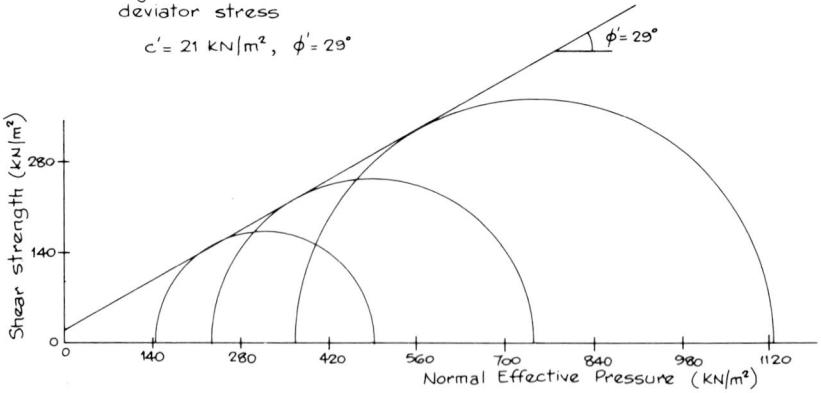

Analysis based on maximum deviator stress

$c' = 21$ kN/m², $\phi' = 29°$

$\phi' = 29°$

Shear strength (kN/m²)

Normal Effective Pressure (kN/m²)

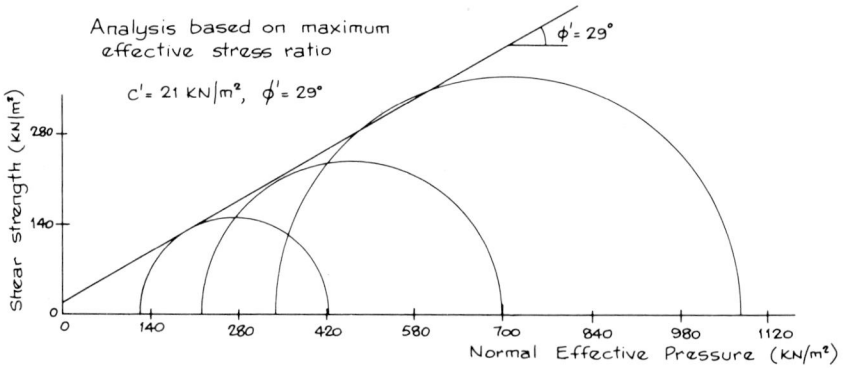

Analysis based on maximum effective stress ratio

$c' = 21$ kN/m², $\phi' = 29°$

$\phi' = 29°$

Shear strength (kN/m²)

Normal Effective Pressure (kN/m²)

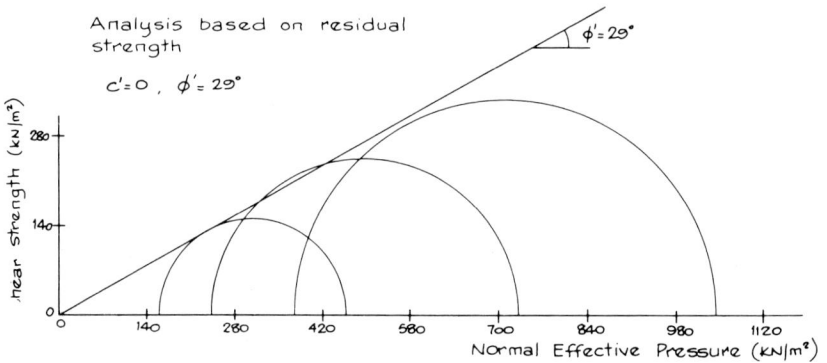

Analysis based on residual strength

$c' = 0$, $\phi' = 29°$

$\phi' = 29°$

Shear strength (kN/m²)

Normal Effective Pressure (kN/m²)

commercial basis as the most sophisticated and versatile shear-strength test. Its versatility is due to the facility of drainage control which allows close correlation to field drainage conditions. The sophistication lies in the fact that pore-water pressures, during undrained testing, can be monitored and volume-change observations, during drained testing, can be made. In addition, field levels of all relevant pressures can be simulated, the applied stresses during testing are principal stresses and tests can be performed under controlled stress or controlled deformation rates.

Other important features are that failure can occur on any plane (whereas in the shear box test the failure plane is predetermined) and naturally occurring features of the soil can be incorporated into the samples under test if such samples are of sufficiently large diameter.

2. The failure plane as defined by the Mohr's circle may not be the failure plane in practice. It is only so if the assumptions upon which the theory is based are valid. With sands, the inclination of the failure plane to the horizontal can be in error by $5°$, while a smaller error is usually applicable to clays. This means that the failure plane is not, in fact, inclined at $45° + (\varphi/2)$ as indicated by the Mohr's circle plot.

3. Two important differences between the test conditions and probable field conditions are:

(a) In the field the soil is in a state of anisotropic stress, i.e. the vertical pressure is in excess of the lateral pressure ($\sigma_1 > \sigma_3$), and usually it has attained its state of consolidation under similar stress conditions. However, in the test, the sample is consolidated under isotropic stress conditions ($\sigma_1 = \sigma_3$), i.e. equal all-round cell pressure is applied to consolidate the sample before shearing.

(b) In the triaxial cell the sample is axially symmetrical: $\sigma_2 = \sigma_3$ (Fig. 4.35). However, most problems at field scale approximate to a condition of plane strain; the length-to-breadth ratio is large and consequently the problem can be treated as being two-dimensional, made up of a series of similar strips with no stresses assumed to act between these strips. The assumption of this condition also simplifies the analysis of the problem.

The above differences between field and test conditions, however, appear to counterbalance each other in some way, since experience shows that triaxial test results generally correlate well with observed field values of shear strength.

4. Soil mechanics tends to be dominated by shear-strength theory and problems such as bearing capacity and slope stability are generally analysed by a limit analysis: the limiting condition is analysed and an appropriate factor of safety is chosen. The value of this factor of safety against shear failure is chosen on the basis of experience and takes into account what are considered to be acceptable deformations. For instance, following the work of Terzaghi and others, a common factor of safety for bearing-capacity problems is of the order of 3, while for slope-stability problems the factor of safety against shear failure is usually about 1·5 or so.

Drained Triaxial Test.

Results for one of 3 samples tested.

Time (min)	Strain Dial G. (mm) Δℓ	Strain ε	Burette (cc)	Volume Change ΔV (cc)	$\frac{\Delta V}{V_0}$	$1 - \frac{\Delta V}{V_0}$	$\frac{\Delta \ell}{\ell_0}$	$1 - \frac{\Delta \ell}{\ell_0}$	Area $= A_0 \frac{1 - \Delta V/V_0}{1 - \Delta\ell/\ell_0}$ mm²	Stress Dial G (Divs)	Load (kN)	Deviator Stress $(\sigma_1 - \sigma_3)$ kN	DVM (mv)
0	0	.00	3.12	0	0	1	0	1	1134.1	0	0	0	+3.11
20	.304	.004	3.17	.05	.00058	.9994	.0040	.9960	1138.0	13.8	.046	40.42	+3.10
39	.593	.0078	3.27	.15	.00174	.9983	.0078	.9922	1141.1	22.5	.074	64.85	+3.11
58	1.216	.0116	3.38	.26	.0030	.9970	.0160	.9840	1149.1	25	.083	72.23	+3.11
77	1.710	.0154	3.57	.45	.0052	.9948	.0154	.9846	1145.9	28.8	.095	82.90	+3.10
96	1.459	.0192	3.77	.65	.0075	.9925	.0192	.9808	1147.6	31.3	.103	89.75	+3.12
115	1.748	.0250	3.87	.75	.0087	.9913	.0230	.9770	1150.7	32.5	.107	93.00	+3.15
150	2.28	.030	4.10	1.05	.0122	.9878	.0300	.9700	1154.9	37.5	.124	107.37	+3.13
186	2.83	.0372	4.47	1.35	.0157	.9843	.0380	.9620	1160.4	41.3	.136	117.20	+3.10
223	3.39	.0446	4.94	1.82	.0211	.9789	.0446	.9554	1162.0	45.0	.149	128.23	+3.12
255	3.88	.051	5.19	2.07	.0240	.9760	.0511	.9489	1166.5	48.8	.161	138.02	+3.11
289	4.39	.0578	5.46	2.34	.0272	.9728	.0578	.9422	1170.9	62.5	.173	147.75	+3.13
327	4.97	.0654	5.76	2.64	.0306	.9694	.0654	.9346	1176.3	57.5	.190	161.52	+3.11
362	5.50	.0724	6.01	2.89	.0335	.9665	.0724	.9276	1181.7	61.3	.202	170.94	+3.13
397	6.03	.0794	6.26	3.14	.0364	.9636	.0793	.9207	1186.9	63.8	.211	177.77	+3.14
433	6.58	.0866	6.47	3.35	.0389	.9611	.0866	.9134	1193.3	66.3	.219	183.53	+3.10
468	7.11	.0936	6.74	3.62	.0420	.9580	.0936	.9064	1198.7	68.8	.227	189.37	+3.09
503	7.65	.1006	6.94	3.82	.0443	.9567	.1007	.8993	1205.2	71.3	.235	194.99	+3.10
537	8.16	.1074	7.22	4.10	.0476	.9527	.1074	.8926	1210.1	73.8	.244	201.64	+3.11
572	8.69	.1144	7.45	4.33	.0502	.9498	.1143	.8857	1216.2	76.3	.252	207.20	+3.13
608	9.24	.1216	7.61	4.49	.0521	.9479	.1216	.8784	1223.8	78.8	.260	212.46	+3.11 *Failure
642	9.76	.1284	7.77	4.65	.0540	.9459	.1284	.8716	1230.8	78.4	.258	210.43	+3.10

AT FAILURE $\sigma_1 = 352.45$ kN/m²; $\sigma_3 = 140$ kN/m²

Sample A1

Cell Pressure = 140 kN/m²

Initial height of sample = 76 mm = ℓ_0

Initial volume of sample = 86.19 cc = V_0

Initial diameter of sample = 38mm

Rate of Strain = .02% min = .0152 mm/min

Proving Ring : 1 div = 3.3 N

Time to failure = 10 hr 8 min

The determination of a factor of safety requires a confident, sensible prediction of the shear strength. The choice of the appropriate shear-strength parameters thus assumes great importance. These parameters can be obtained from undrained, consolidated undrained, drained or, in appropriate cases, perhaps long-term large-strain tests. The appropriate parameters are as follows:

C_u and φ_u from undrained tests

C_{cu} and φ_{cu} from consolidated undrained tests which, if performed with measurement of pore-water pressures, can lead to effective stress parameters C' and φ'

C_d and φ_d from drained tests

C_r and φ_r from large-strain tests and termed 'residual shear strength parameters'.

All but the last-named set of parameters are commonly found by triaxial testing with appropriate drainage conditions. The residual-strength parameters are usually determined from ring shear tests (Section 4.3.2).

Any practical problem then requires satisfactory answers to the following questions:

(i)   How are these parameters related to the problem?

(ii)   Should the analysis be in terms of effective stress or total stress?

(iii) Which condition gives the critical design case? (Usually the lowest factor of safety against shear failure during the lifetime of the proposed works.)

(iv) How might the pore water pressure (and consequently the shear strength) vary with time?

(v)   How might the drainage conditions affect the problem?

(vi) What is the expected rate of loading?

(vii) How will the history of the soil affect the shear strength? (For example, is it over-consolidated, normally-consolidated, has it been subject to large strain, is it varved or seasonly layered, and so on.)

Such questions illustrate that the application of shear-strength theory and test results to practical problems is relatively complex and needs careful thought before deciding and planning a site investigation and a correspond-

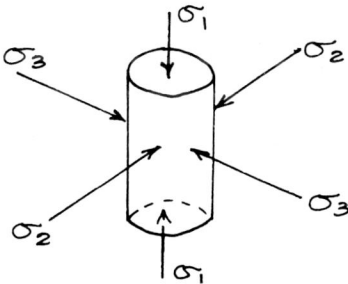

Fig. 4.35 Axially symmetrical sample

ing test programme. Broadly speaking, the following general applications of tests to problems can be made.

For the immediate (short-term) stability problem associated with impervious (slow-draining) clays, *undrained* tests are appropriate so long as no change in moisture content will take place at the field scale. Note that clays which have fissures, silt layering, organic intrusions or a complex structure are likely to be fairly quick-draining so that problems arise in applying undrained test results.

If drainage is allowed, one may consider using *consolidated undrained* tests with observations of pore-water pressure to give C' and φ' as the effective stress parameters. Here it is important to realise that the validity of the shear-strength calculation is largely dependent on the accuracy of pore-water pressure observations, so that field measurement (by piezometer) may be necessary to verify laboratory or other predictions.

In some cases, rather than using C' and φ' values obtained from consolidated undrained tests, it is preferable to perform *drained* tests to give C_d and φ_d. This is because, in drained tests, volume change takes place during shear (due to the expulsion of pore water), indicating similar volume change at field scale. The strength parameters C' and φ' are, similarly, useful in stability problems involving hydrostatic conditions or steady seepage conditions.

Residual-strength tests are not easy to apply; however, where there is definite knowledge that previous slip has occurred, their use is essential. The difficulty lies in deciding to what extent the state of the soil approximates to residual strength parameters. Figure 4.36 illustrates the residual and peak-strength values.

5. Table 4.1 indicates some typical field problems together with suggested appropriate shear-strength parameters obtained from the corresponding tests.

6. It should be noted that three important assumptions are implicit in the equation describing the Mohr–Coulomb failure line:

$$S = C + \sigma_n \tan \varphi$$

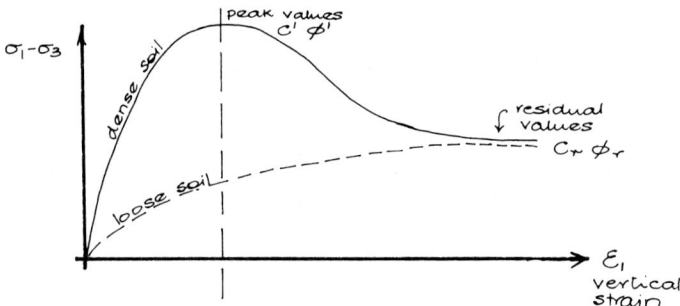

Fig. 4.36 Values of residual strength and peak strength

Table 4.1   Shear-strength parameters for typical field problems

Type of problem	Load	Critical period	Analysis and comment
Foundations on saturated clay. Passive earth pressures on retaining walls	Positive	End of construction	Total stress C_u, $\varphi_u = 0$. Experience shows that such analysis gives acceptable solutions
Earth dam construction Embankment fill. Both can involve staged construction (i.e. several loading periods)	Positive	End of construction	Effective stress C', φ'. Staged construction gives some dissipation of excess pore-water pressure. Best to measure pore-water pressure at field scale and check the factor of safety at each stage
Earth pressures on walls backfilled with partially saturated material ($\varphi_u \neq 0$)	Positive	End of construction *or* long-term	Total stress C_u, $\varphi_u = 0$. Effective stress C', φ'. Seepage pore-water pressures, if any
Active pressure on driven and dredged sheet pile wall. drive — dredge	Negative	Usually long-term	Effective stress C', φ'. Pore-water pressure from least favourable position of water table behind the wall
Permanent cuts. Stability of natural slopes	Negative	Long-term	Effective stress C', φ' or preferably C_d, φ'_d. Long-term pore-water pressures from steady seepage or static condition. C' is not reliable and is often taken as zero. For some fissured over-consolidated clays use residual parameters $C_r = 0$, φ_r
Temporary excavations. Slope stability. Base heave of *intact* clays	Negative	During construction	Total stress C_u, $\varphi_u = 0$. C_u preferably measured in unloading type of triaxial test
Temporary excavations. Slope stability. Base heave of *non-intact* clays	Negative	During construction	Effective stress $C'\varphi'$. Quick drainage makes an undrained analysis unreliable. Often necessary to estimate pore-water pressures which can be difficult

It is assumed

(a) that the equation is valid for any value σ_2 = intermediate principal stress (whereas, in the triaxial test, $\sigma_2 = \sigma_3$ in all cases);

(b) that C and φ have the same value for every section through a soil mass (this assumption is not necessarily correct);

(c) that C and φ are independent of the stress states preceding failure.

7. The correlation between the types of triaxial test in terms of total stress and effective stress is shown in Fig. 4.37. It should be noted that during the unconfined compression test pore-water pressure within the sample is negative (i.e. suction).

8. As mentioned previously, the choice of fluid for application of the cell pressure can be important. For most practical purposes water is eminently suitable but there is evidence that, in tests of long duration, water tends to permeate through the membrane. Some success has been achieved in long-term testing by using light oils or glycerine. Although such fluids are more difficult to handle, they can be of benefit in terms of test quality.

9. Triaxial test results can be presented in various ways. Datasheets Nos 11–13 illustrate one way of presenting undrained, consolidated undrained, and drained test results.

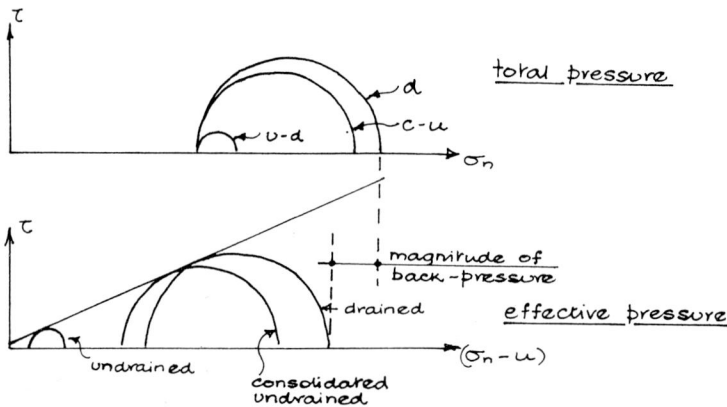

Fig. 4.37

4.6 Determination of Skempton's Pore Pressure Coefficients

The now well-established pore pressure coefficients proposed by Skempton[12] are of importance in geotechnical engineering and are used extensively in applying shear strength characteristics to practical problems at field scale. They are particularly useful in prediction of pore-water pressure behaviour and, since pore-water pressure is related to shear strength, in the prediction of factors of safety with respect of shear strength.

Skempton proposed two coefficients to express how pore pressures responded to changes in total stress under *undrained* conditions.

4.6.1 The Coefficient B

The coefficient (B) is defined as follows

$$B = \frac{\Delta u_3}{\Delta \sigma_3}$$

where $\Delta \sigma_3$ = increase in total stress in each direction
    Δu_3 = immediate increase in pore-water pressure resulting from stress.

Assuming that the compressibility of the solid particles is zero. The compressibility of the pore water is negligible in a fully saturated soil (water only present in the voids) and B takes the value of 1.

Thus if a sample of saturated soil is subject to a pressure increase all round (see Fig. 4.38) $(\Delta \sigma_3)$ of say 50 kN/m^2 the corresponding change in

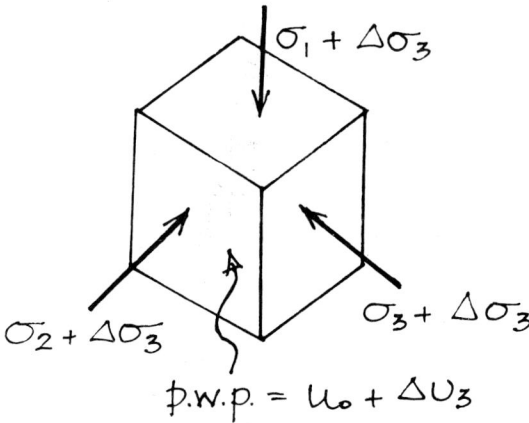

Fig. 4.38

pore-water pressure will be given by

$$B = 1 = \frac{\Delta u_3}{\Delta \sigma_3} = \frac{\Delta u_3}{50}$$

whence $\Delta u_3 = 50$ kN/m^2.

For partially saturated soils, the compressibility of the pore fluid is high (air and water in the voids) and B is less than 1. A typical variation of B with degree of saturation of the soil is given in Fig. 4.39. It should be noted that there is no unique relationship between the coefficient 'B and the degree of saturation (s) but that it will vary for each soil type.

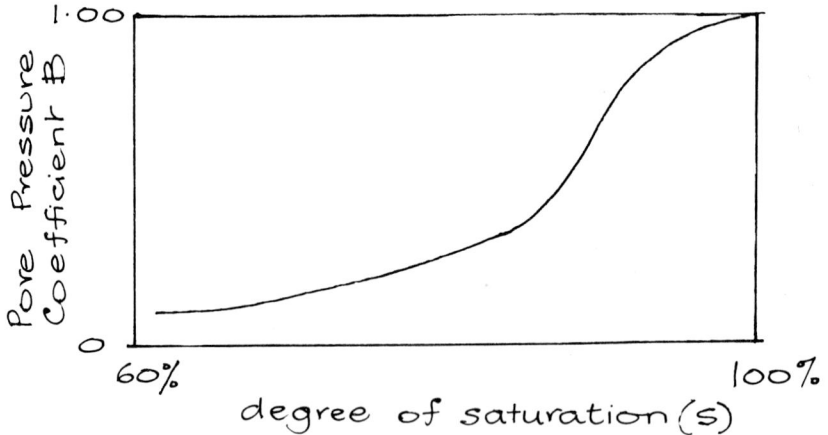

Fig. 4.39

Thus, for example, if a soil, saturated to 90%, with $B = 0\cdot60$ say, is subject to an equal all-round pressure increase ($\Delta\sigma_3$) of 50 kN/m², the corresponding change in pore-water pressure will be

$$\Delta u_3 = B(\Delta\sigma_3) = 0\cdot6(50) = 30\ kN/m^2$$

assuming that no drainage takes place.

The value of the coefficient B is measured in the triaxial test, usually as the cell pressure is increased.

A sample is set up under an observed cell pressure and the pore-water pressure is measured. The cell pressure can then be increased, under conditions of no drainage, and the *change* in pore-water pressure can be observed. By using the expression

$$\Delta u_3 = B(\Delta\sigma_3)$$

and observing values of $\Delta\sigma_3$ and Δu_3, then B can be calculated.

In this way, the cell pressure can be incrementally increased and B observed at each stage, for each cell pressure range. It should be noted that at high values of cell pressure it is possible to achieve a state of full saturation, because air in the void space will go into solution and the soil will become more saturated. Conversely, if the cell pressure is decreased air can come out of solution and the value of B will decrease.

By using this technique it is possible to determine the level of back-pressure to be used for testing to ensure that full saturation is maintained. This will be especially important if the sample is to be sheared undrained and a reduction in pore-water pressure is expected.

4.6.2 The Coefficient A

If the major principal stress *only* is increased (see Fig. 4.40) ($\Delta\sigma_1$) then a second change in pore-water pressure will occur so long as undrained conditions are maintained.

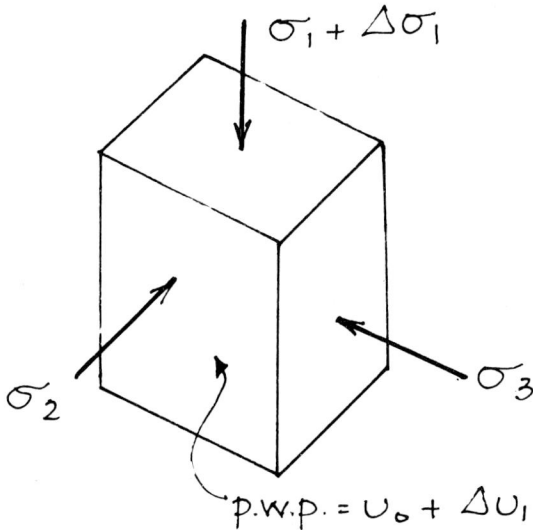

Fig. 4.40

Skempton related this change in pore-water pressure (Δu_1) to the change in major principal stress by the expression

$$\Delta u_1 = A.B.(\Delta\sigma_1)$$

where A is a pore pressure to be determined experimentally. Sometimes AB is re-written as $\bar{A}$.

For the particular case of a saturated soil with $B = 1$, then

$$\Delta u_1 = A(\Delta\sigma_1)$$

Once again it is imperative that changes in both total stress and pore-water pressure occur under undrained conditions.

During the shearing stage of a triaxial test, the total vertical stress σ_1 is continually increased, either steadily or incrementally, thus $\Delta\sigma_1$ is continually varying. This means that A will vary throughout the test and, usually, A_f, the value of A at failure is quoted in results and reports.

The variation of A during a test will also depend upon the history of the soil, i.e. over-consolidated or normally consolidated, and upon the soil type. This is one important variation between the values of the coefficients A and B.

For normally consolidated soils typical values of A_f may be in the range 0·5 to 1·0. In some clays very high pore-water pressure changes may be observed and A can be greater than 1·0.

For over-consolidated soils, negative pore-water pressure changes may be generated and A_f can have negative values perhaps of the order of -0.5.

A typical variation of A_f with over-consolidation ratio is given in Fig. 4.41.

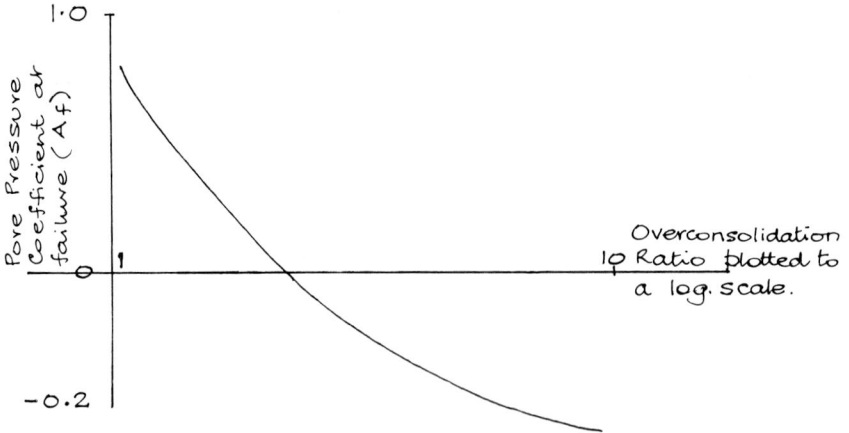

Fig. 4.41

4.6.3 The Coefficient $\bar{B}$

The pore pressure coefficients A and B are often combined to give the total pore-water pressure change (Δu) due to both the all-round total pressure increase and the axial total stress increase.

$$\Delta u = B\Delta\sigma_3 + AB(\Delta\sigma_1 - \Delta\sigma_3)$$

$$\Delta u = B\left[\Delta\sigma_3 + A(\Delta\sigma_1 - \Delta\sigma_3)\right]$$

Dividing throughout by $\Delta\sigma_1$ then

$$\frac{\Delta u}{\Delta\sigma_1} = B\left[\frac{\Delta\sigma_3}{\Delta\sigma_1} + A\left(1 - \frac{\Delta\sigma_3}{\Delta\sigma_1}\right)\right]$$

which can be re-written as

$$\frac{\Delta u}{\Delta\sigma_1} = B\left[1 - (1 - A)\left(1 - \frac{\Delta\sigma_3}{\Delta\sigma_1}\right)\right] = \bar{B}$$

thus introducing a further coefficient $\bar{B}$ which has wide practical applications.

CHAPTER 5

Consolidation Tests

Determination of consolidation characteristics of saturated clays using

   (*a*) Casagrande oedometer test
   (*b*) Rowe consolidation cell.

5.1 Introduction

When a saturated clay is subject to an increment of total pressure Δp, a corresponding excess pore-water pressure Δu is set up within the void space of the soil. The magnitude of Δu is numerically equal to Δp initially, assuming that the pore water is relatively incompressible. This is adequately explained by Terzaghi's spring and piston analogy.

At the instant of loading, assuming that the drain is closed, any applied load is carried totally by the water, since the spring cannot take load if it is not allowed to deform.

As can be seen from Fig. 5.1, when drainage is allowed, water seeps out of the container, the pressure drops, the spring deforms (thus taking load) and as time proceeds this process continues until ultimately all the applied load is carried by the spring and the water pressure has fallen to the value recorded before the application of the load increment.

The process described by the model is similar to the process that takes place when saturated clays are loaded, the soil structure replacing the spring and the pore water in the voids of the soil replacing the water within the container. The rate of flow of water from the soil depends upon the permeability of the soil and the length of the drainage path that particles of water follow. The resultant settlement of the piston in the model is replaced by the settlement of the surface of the clay layer. Thus two important conclusions can be drawn: (i) the settlement is a function of time, and (ii) there is a continual decrease in the pore-water pressure while the settlement proceeds, and therefore three quantities are of interest to the foundation engineer:

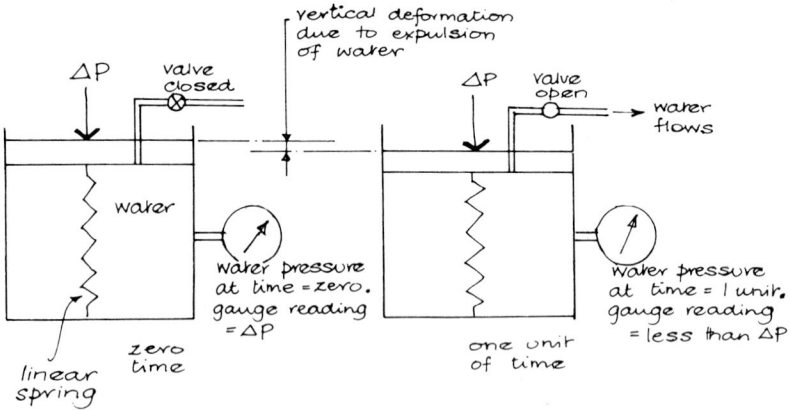

Fig. 5.1 Terzaghi's spring and piston analogy

The magnitude of the settlement
The rate at which this settlement occurs
The variation in shear strength occurring during settlement because of the changing pore-water pressure.

Consolidation theory is therefore directly relevant to settlement predictions and stability analysis involving shear strength. In practice, most consolidation problems involve the three-dimensional flow of water and associated three-dimensional strain. However, three-dimensional theory introduces mathematical complexities and has limited use and so most work is based upon Terzaghi's one-dimensional consolidation theory involving one-dimensional flow of water and one-dimensional strain. Problems involving two-dimensional flow of water to sand drains are usually analysed using Barron's theory.

For the one-dimensional case, Terzaghi formulated a partial differential equation relating excess pore-water pressure, time and depth of an element of soil subject to a suddenly applied, uniformly distributed load over an infinite area of the soil (Fig. 5.2). The governing partial differential equation is

$$\frac{\partial u}{\partial t} = C_v \frac{\partial^2 u}{\partial z^2}$$

where u = excess pore-water pressure, t = time elapsed since loading and C_v = coefficient of consolidation:

$$C_v = \frac{K}{\gamma_w m_v} \text{ m}^2/\text{year}$$

where γ_w = unit weight of water (kN/m^3)
K = vertical coefficient of permeability of the soil (m/yr)
m_v = coefficient of volume compressibility, volume change, volume decrease (m^2/kN).

Terzaghi's solution to this partial differential equation was then expressed in dimensionless form in terms of two quantities, T_v and U_v:

$$T_v = \text{Terzaghi time factor} = \frac{C_v t}{d^2}$$

where d = drainage path length.

$$U_v = \text{average degree of consolidation} = \frac{S_t}{S_{ult}} = \frac{\text{settlement at time } t}{\text{ultimate settlement}}$$

Published tables of U_v and the related T_v values are available for a variety of distributions of excess pore-water pressure with depth in the clay layer at the *instant of loading* (before drainage).

The drainage path length, d, as used in calculation of the dimensionless time factor T_v, is a function of the drainage facilities adjacent to the clay layer. Usually, either $d = H$ with one-way drainage where an impervious bed is adjacent to one horizontal surface of the clay, or $d = H/2$ with two-way drainage where facility for drainage exists adjacent to both horizontal boundaries of the clay.

For the case of *uniform* excess pore-water pressure distribution with depth, the corresponding U_v–T_v relationship is:

U_v (%)	10	20	30	40	50	60	70	80	90
T_v	0·008	0·031	0·071	0·126	0·197	0·287	0·403	0·567	0·848

The function of vertical consolidation tests performed in the laboratory is to judge if the use of the Terzaghi consolidation theory is justifiable in civil engineering practice. This judgement is performed by predicting settlement values and the rates at which these settlements will occur.

Settlement of a saturated clay layer is usually the result of void ratio reductions (due to the expulsion of water from within the voids). It has been

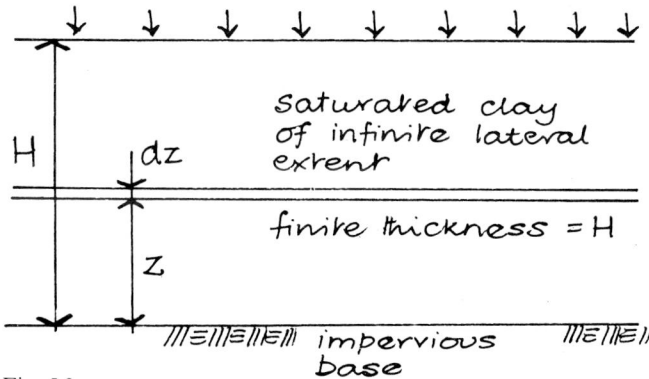

Fig. 5.2

shown that the void ratio of a soil is related to the effective pressure and this fact is used in predicting amounts of settlement.

For one-dimensional (vertical) settlement it can be shown that

$$\frac{\Delta H}{H_0} = \frac{\Delta e}{1 + e_0} \qquad (5.1)$$

where ΔH = settlement
      H_0 = initial thickness
      Δe = change in void ratio
      e_0 = initial void ratio.

5.1.1   Normally-consolidated and Over-consolidated Soils

Two types of soil are commonly encountered in practice:

(i) *Normally-consolidated soils* are soils whose existing effective pressure is the maximum effective pressure to which they have been subjected in their history.

(ii) *Over-consolidated soils* are soils whose existing effective pressure is less than a previous maximum effective pressure to which they were once subjected. Over-consolidation is thus the result of a reduction in total pressure or an increase in pore-water pressure at some time in the history of a soil. Common causes of over-consolidation are excavation, preloading, erosion and variation in ground-water level.

The over-consolidation ratio is defined as

$$\frac{\text{maximum previous effective pressure}}{\text{existing effective pressure}}$$

This ratio is, of course, unity for normally-consolidated soils and greater than unity for over-consolidated soils.

Methods for the determination of the maximum previous effective pressure (often called the preconsolidation pressure) have been suggested by Casagrande [10] and Schmertmann.[11]

The relevance of this distinction between types of soil is that a normally-consolidated soil subject to an effective pressure increase settles more than does an over-consolidated soil subject to an identical pressure increase. This can be seen from the two plots of void ratio *e* versus effective pressure *p* given in Fig. 5.3. These curves reveal another important characteristic of consolidation: that it is not a wholly reversible process, i.e. only a part of the compression of a soil is recoverable on removal of load. This is because, during compression, some rearrangement of the particles takes place in addition to the expulsion of water from the voids; on removal of the load, when the soil takes up water, not all of the 're-packed' soil particles can

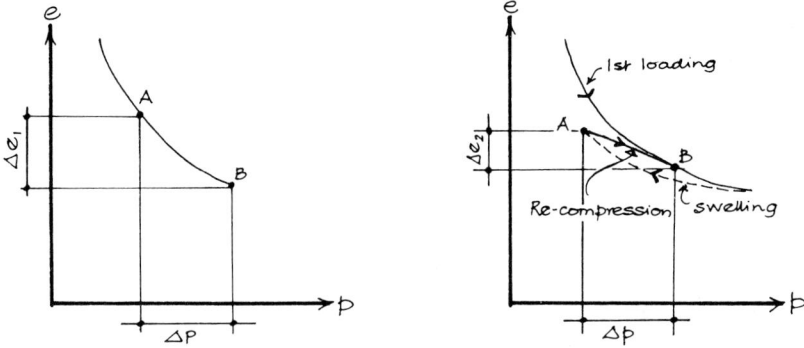

Fig. 5.3 Void ratio e related to effective pressure p
(a) normally-consolidated soil   (b) over-consolidated soil
A = initial void ratio and corresponding pressure
B = final void ratio and corresponding pressure
Δp = identical increment for each soil
$\Delta e_1 > \Delta e_2$ = changes in void ratio

regain their original positions. The amount of this 'irrecoverable compression' varies from soil to soil, depending very much upon the soil type and original structure.

If the void ratio–effective pressure curves are replotted as e versus $\log_{10} p$, the results shown in Fig. 5.4 are obtained.

This gives an instant indication of whether a soil is over-consolidated (predominantly curved e versus $\log_{10}p$) or normally-consolidated (linear e versus $\log_{10}p$).

The gradient of the e versus $\log_{10}p$ plot for normally-consolidated soils is given the symbol C_c = the compression index:

$$C_c = \frac{e_o - e_f}{\log_{10}(p_f) - \log_{10}(p_o)} = \frac{\Delta e}{\log_{10}(p_f/p_o)}$$

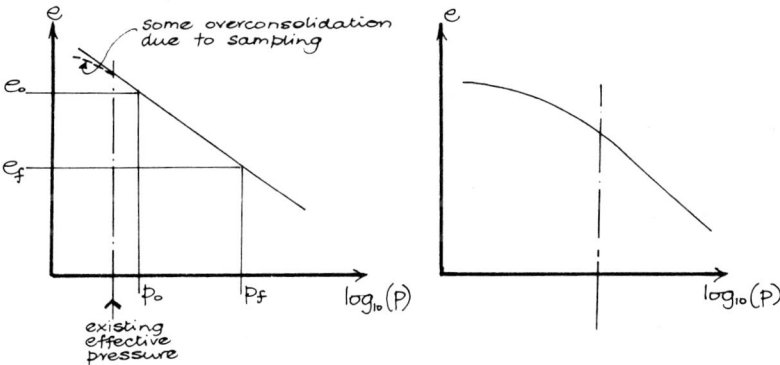

Fig. 5.4 Void ratio e related to effective pressure $\log_{10} p$
(a) normally-consolidated soil   (b) over-consolidated soil

thus

$$\Delta e = C_c[\log_{10}(p_f/p_o)]$$

for a normally consolidated soil.

Using eq. (5.1),

$$\frac{\Delta H}{H_o} = \frac{\Delta e}{1 + e_o} = \frac{C_c \log_{10}(p_f/p_o)}{1 + e_o}$$

or

$$\Delta H = \frac{C_c H_o}{1 + e_o} \log_{10} \frac{p_f}{p_o} \tag{5.2}$$

Equation (5.2) gives the ultimate or final consolidation settlement of a normally-consolidated clay.

The compression index is inappropriate to over-consolidated soils so usually ultimate settlement predictions for such soils are based upon m_v values. m_v, the coefficient of volume change, is defined as the change in volume per unit volume per unit increase in pressure.

Consider a soil sample of constant unit volume of solids (i.e. the volume of solids is one cubic unit of volume):

$$\text{Initial volume} = \underset{\text{(solid)}}{1} + \underset{\text{(water)}}{e_o}$$

Any change in volume will be the result of change in void ratio, i.e. the volume $(1 + e_o)$ reduces by Δe under a pressure increase of Δp. Thus

$$m_v = \frac{\Delta e}{(1 + e_o)/\Delta p} = \frac{\text{(change in volume)}}{\text{(original volume)/(increase in pressure)}}$$

or

$$\frac{\Delta e}{1 + e_o} = m_v \Delta p$$

Substituting this value in eq. (5.1),

$$\frac{\Delta H}{H_o} = \frac{\Delta e}{1 + e_o} = m_v \Delta p$$

or

$$\Delta H = m_v \Delta p H_o \tag{5.3}$$

Equation (5.3) is used to predict ultimate settlements in over-consolidated deposits.

It must be noted that

$$m_v = \frac{\Delta e}{\Delta p}\left(\frac{1}{1 + e_o}\right)$$

or

$$m_v = \left\{\begin{matrix} \text{gradient of} \\ e\ v.\ p\ \text{curve} \end{matrix}\right\} \times \frac{1}{1 + e_o}$$

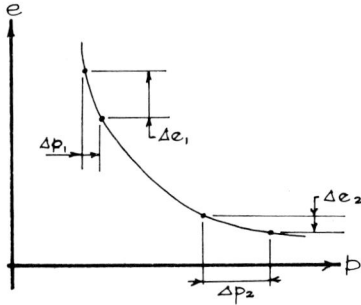

Fig. 5.5

This gradient is defined as $a_v = \Delta e / \Delta p = $ coefficient of compressibility, and is a function of the pressure range. That is,

$$\frac{\Delta e_1}{\Delta p_1} \neq \frac{\Delta e_2}{\Delta p_2}$$

Specific values of m_v and a_v must therefore be allotted to each range of effective pressures and the value of m_v used in ultimate settlement predictions *must* correspond to the pressure range predicted for the full-scale field problem. m_v is commonly expressed in m^2/kN.

5.1.2 Prediction of Settlement Rate

To predict the rates at which settlement will occur, use is commonly made of the Terzaghi theory and this requires the determination of the coefficient of consolidation (C_v). Two methods are given below. Both methods involve 'curve-fitting' techniques; the reasoning behind the use of such methods is explained in most soil mechanics textbooks.

(a) *Taylor and Merchant Method*
Observations of settlement and time are recorded and a plot of settlement (s) versus square root of time (t) is made.

For 90% consolidation, assuming uniform distribution of excess porewater pressure with depth, $T_v = 0.848$ from the Terzaghi time factor values. Now

$$T_v = \frac{C_v t}{d^2}$$

or

$$C_v = \frac{T_v d^2}{t}$$

Therefore if

$$U_v = 90\%$$

then

$$C_v = \frac{0.848 \times d^2}{t_{90}}$$

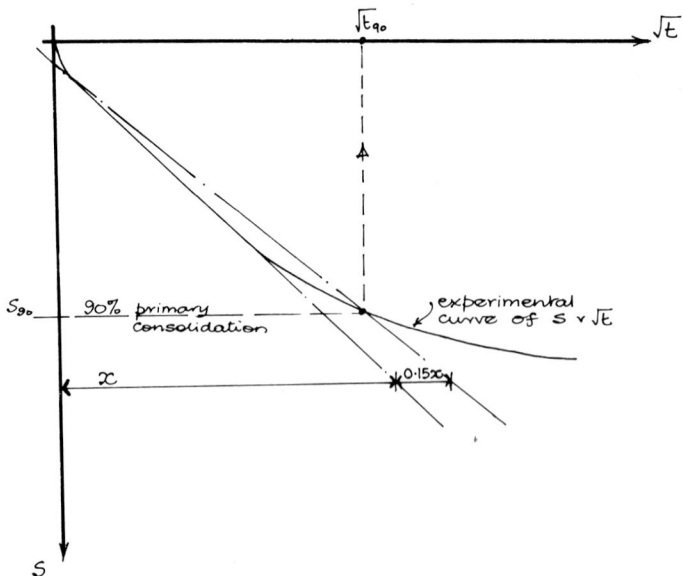

Fig. 5.6 Prediction of settlement rate: Taylor and Merchant method

t_{90} is found from Fig. 5.6 and $d =$ the drainage path length appropriate to the laboratory test: with 'one-way' drainage $d =$ sample thickness; with 'two-way' drainage $d =$ (sample thickness)/2. Consequently C_v can be determined, noting that

(i) C_v often varies with varying pressure increments;

(ii) C_v is expressed usually in $m^2/year$.

(b) Casagrande Method
Observations of settlement and time are plotted as settlement versus the logarithm of time and the characteristic shape shown in Fig. 5.7 is produced.

From the tabulated Terzaghi U_v/T_v relationships, at 50% consolidation $(U_v = 50\%)$ the corresponding time factor $T_v = 0.197$, assuming initial uniform distribution of excess pore-water pressure with depth.

$$T_v = \frac{C_v t}{d^2}$$

and also

$$T_v = 0.197 = \frac{C_v t_{50}}{d^2}$$

therefore

$$C_v = \frac{0.197\, d^2}{t_{50}}$$

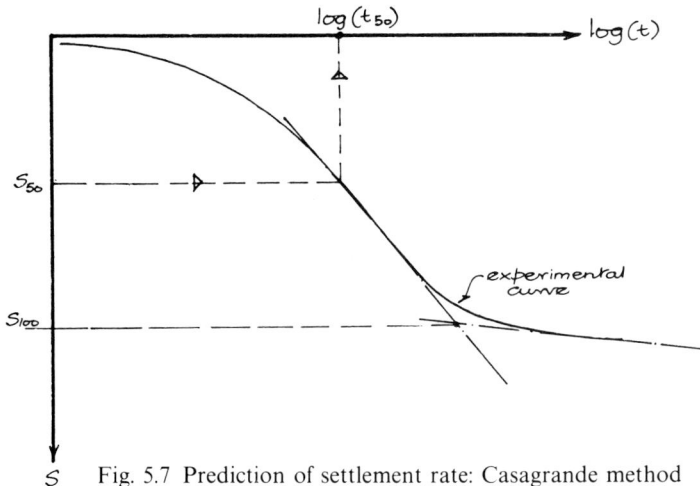

Fig. 5.7 Prediction of settlement rate: Casagrande method

t_{50} is determined by the construction shown in Fig. 5.7 and d is the appropriate drainage path length.

Types of Settlement
In both of the methods outlined, three distinct and often interrelated types of settlement are evident. These are usually termed

immediate settlement
primary consolidation
secondary compression.

These three characteristic settlements can be broadly identified in both methods as follows.

(a) *Taylor and Merchant method*. The immediate settlement occurs fairly rapidly and is usually neglected when considering rates of consolidation in the long term. The point where primary consolidation is deemed to begin is obtained by producing the straight line portion of the S versus t plot back to the settlement axis (point a, Fig. 5.8).

The point where the primary consolidation is assumed to finish is found by setting $ac = ab/0.9$ on the settlement axis and then producing the point c horizontally to cut the experimental curve.

Any further settlement below the horizontal line cd is said to be secondary compression. It is worthy of note that the amount of secondary compression can, with some types of soils, be of a significant magnitude. Also it must be realised that secondary compression does not, as Fig. 5.8 may imply, begin only after the primary consolidation is complete. Secondary compression occurs simultaneously with primary consolidation and starts immediately after loading.

Its cause has been subject to wide investigation and it is now generally accepted that often it is associated with the viscous resistance of soil particle movement caused by the viscosity of the adsorbed water layers surrounding the individual solid particles. Secondary compression leads to evidence that the void ratio is not uniquely related to effective pressure but that it is also a function of time.

(b) The *Casagrande method* similarly identifies these three types of settlement (Fig. 5.9). Since in the early stages of consolidation the time factor is proportional to the square of the average degree of consolidation,

$$T_v \propto (U_v)^2 \quad \text{and} \quad \frac{C_v t}{d^2} \propto \left(\frac{S_t}{\text{ultimate } S}\right)^2$$

Thus at two early values of time, t_1 and t_2, assuming that C_v, d and ultimate settlement are constant,

$$\frac{t_1}{t_2} = \left(\frac{S_1}{S_2}\right)^2$$

Therefore if

$$S_2 = 2 \times S_1$$

then

$$\frac{t_1}{t_2} = (\tfrac{1}{2})^2$$

thus

$$t_2 = 4t_1$$

To estimate the datum for primary consolidation settlements, two times are chosen on the early part of the curve such that $t_2 = 4 \times t_1$; the settlements from this datum will therefore be in the ratio of two to one. A

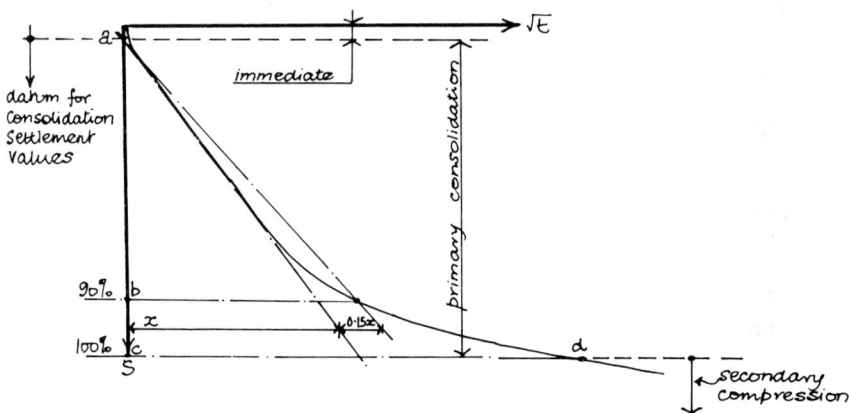

Fig. 5.8 Identification of types of settlement: Taylor and Merchant method

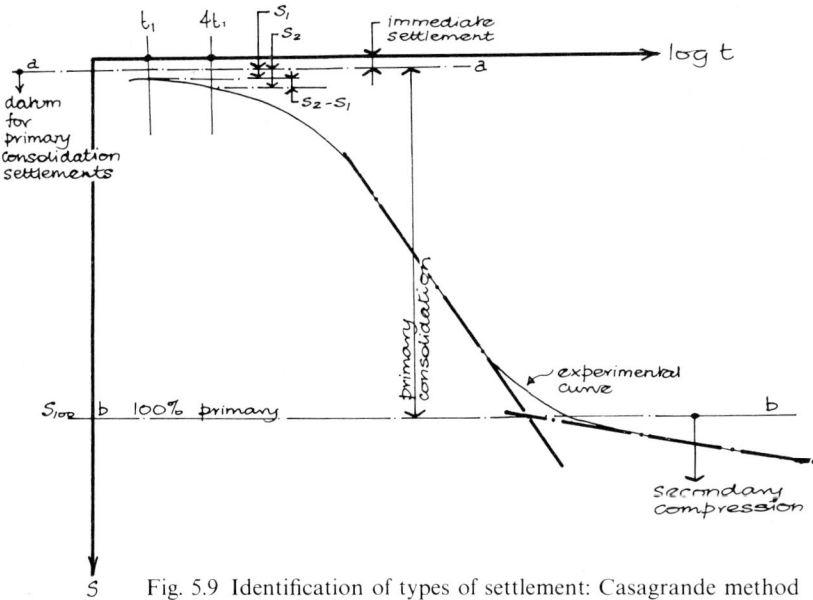

Fig. 5.9 Identification of types of settlement: Casagrande method

simple construction then results. Choose t_1, set out $4 \times t_1$; the settlement *between* these two times $= S_2 - S_1$ and, because $2S_1 = S_2$, setting out a settlement of this difference above $a - a$ gives the datum for primary consolidation settlements.

Any settlements below the line $b - b$ are attributed to secondary compression effects. If consolidation tests are sustained for a significant length of time, estimates of C_α, the coefficient of secondary compression, may be made. It should be noted that this secondary tail to the S v. $\log t$ curve, although apparently of constant gradient in the early stages, may develop marked curvature at larger time values.

The coefficient of secondary consolidation C_α is defined as change in void ratio per log cycle of time. Since

$$\frac{\Delta e}{1 + e_0} = \frac{\Delta H}{H_0}$$

thus

$$C_\alpha = \frac{\Delta e}{(\text{log cycle of time})} = \frac{1 + e_0}{H_0} \times \frac{\Delta H}{(\text{log cycle of time})}$$

where e_0 = initial void ratio, H_0 = initial thickness, ΔH = settlement, and $\Delta H/(\text{log cycle of time})$ = gradient of the secondary tail produced in Casagrande's construction.

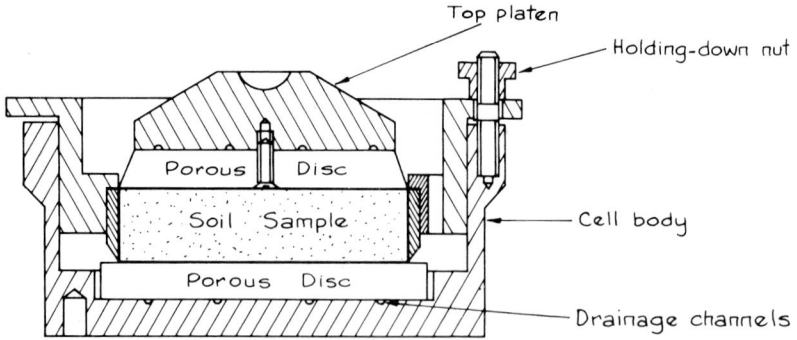

Fig. 5.10 Casagrande oedometer: section

5.2. Casagrande Oedometer Test

The Casagrande type of oedometer is shown in section in Fig. 5.10 and the loading device is illustrated in Fig. 5.11. Usually the sample size is 76 mm diameter by 19 mm thick and it is prepared directly on extrusion from a U4 or similar sample tube.

Initially the porous stones or discs which are placed in contact with the sample faces are saturated by boiling in de-aired water. The physical dimensions of the cutting-ring, which is used as the consolidation test ring, are checked accurately using a micrometer and its mass is determined.

The cutting-ring is then carefully pushed into a specimen of soil about 100 mm diameter by, say, 40 mm thick and the surplus soil is carefully trimmed. It is important to ensure that the two surfaces of the soil which come into contact with the porous stones are truly plane and these surfaces are best trimmed using a cheese wire. From the resultant cuttings at least three initial

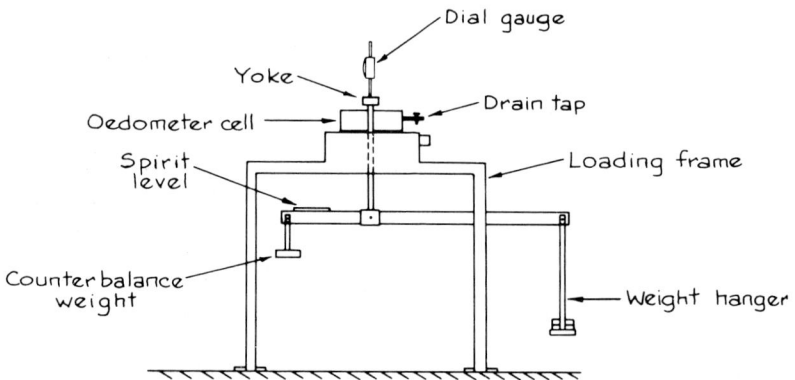

Fig. 5.11 Casagrande oedometer: loading arrangement

moisture-content determinations should be made, and appropriate specific gravity (G_s) values of the solid particles obtained. The cutting-ring and con-tained sample are then weighed and the unit weight is determined.

The lower saturated porous disc is placed on the base of the unit and the water level is raised above the disc. The sample is carefully placed on the lower porous disc; the saturated upper disc and loading platen are then very carefully added. The whole unit is now placed on the pedestal in the loading arrangement, the lever arm is supported and the dial gauge, which seats on the ball bearing, is securely fastened and checked to see that sufficient ver-tical travel is available.

A predetermined loading sequence is decided. Usually the load is increased every 24 hours in the following pattern:

Increment number	1	2	3	4	5	6	7
Time from start of test (h)	0	24	48	72	96	120	144
Pressure on sample (kN/m²)	50	100	200	400	800	200	10
	←		loading		→	unloading	

It is worth noting that with some soils, particularly those that have been lightly loaded in their history, it may be necessary for the pressure increments to start at, say, 25 kN/m². The above values should be regarded purely as 'usual' values, since each type of soil, its location and the function of the test must be considered individually before deciding the pressure-increment sequence. Also, most oedometers are fitted with a lever mechan-ism and compensating arm so that the load applied to the sample is a multiple (often × 10 or × 11) of that applied to the loading platform.

Observations of dial-gauge reading and time from instant of loading are recorded and usually such observations are taken at the following intervals:

$$\text{Time (min)} = 0, \tfrac{1}{4}, 1, 2\tfrac{1}{4}, 4, 6\tfrac{1}{4}, 9, 12\tfrac{1}{4}, 16, 20\tfrac{1}{4}, 25, \ldots, 100$$

and then at suitable intervals up to 24 hours. The purpose of such time intervals is that their corresponding $\sqrt{\text{time}}$ values are $0, \tfrac{1}{2}, 1, 1\tfrac{1}{2}, 2, 2\tfrac{1}{2}, 3, 3\tfrac{1}{2}$, 4 and so on. Settlements will then have been recorded at intervals of time to give an even distribution of points on the graph used in Taylor and Merchant's method for determining the coefficient of consolidation (C_v).

After 24 hours, before applying the next pressure increment, a dial-gauge reading and appropriate time should be recorded. It is also useful to take an observation of the depth from the top of the loading platen to a fixed point on the cell unit (Fig. 5.12). This measurement should be recorded accurately by using a depth micrometer or similar device and can be taken at, say, four points around the cell. Thus, referring to Fig. 5.12, $(a + b) - (c + d) = e$ = thickness of the sample; assuming two-way drainage, $e = 2 \times$ the drainage-path length at any stage. The dimension $e/2$, being a measure of the drainage-path length, is of course used in the determination of C_v from

Terzaghi's time factor in both the Casagrande and the Taylor and Merchant curve-fitting methods.

Throughout the week-long testing period the water levels in the unit and small tank should be maintained so that no desiccation of the sample occurs and so that sufficient water is available to be taken up by the soil when it swells during the unloading sequence.

Upon completion of the swelling process, the sample is dried of surface water immediately and the whole sample is weighed and placed in an oven at 110° C for 24 hours in order to allow a final moisture-content determination. Typical results with appropriate curve-fitting methods are given in Datasheet No. 14.

As in all accurate consolidation testing, the tests should be carried out at constant temperature in a temperature-controlled room. If this is not possible, a record of temperature during testing should be maintained; BS 1377:1975 gives a graph that allows correction then to be made to observed C_v values.

5.2.1   Summary

1. The test provides a useful means of assessing consolidation characteristics of a wide range of clay soils which behave largely in accordance with the Terzaghi theory.

2. The cost of the testing and associated sampling is comparatively low and a variety of samples can be tested simultaneously in a commercial laboratory.

3. The testing time is short because of the thickness of the sample, two-way drainage and small drainage-path lengths.

4. However, it must be noted that there are several major drawbacks to the apparatus: for example, consolidation is concerned with the dissipation

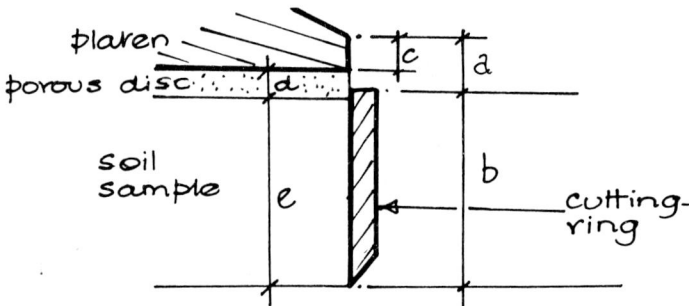

Fig. 5.12 a = depth observed by depth micrometer after each load increment

b = depth of cutting-ring   Note: dimensions b, c and d
c = thickness of load platen   recorded accurately from
d = thickness of top porous disc   micrometer readings

Casagrande Oedometer Test.

Borehole : 4B7   Sample : 1A   2m – 2.5m depth.

Mass of Oed. Ring (g)	Ring + Wet Sample (g)	Ring + Dry Sample (g)	Initial Moisture Content (%)	G_s	Initial Void Ratio e_o
91·8	279·9	252·1	17·3	2·70	0·467

Dial Gauge Readings (mm × 10²)

Time mins.	Initial consolidation pressure (P₀) = 50 kN/m² P_f = 100 kN/m²	P_f = 200 kN/m²	P_f = 400 kN/m²	√Time
0	90·00	70·40	45·60	0
¼	86·00	68·25	43·00	0·5
½	85·25	67·00	42·00	0·71
1	84·75	65·75	40·50	1
1½	84·25	64·75	39·25	1·22
2	84·00	64·00	38·25	1·41
3	83·25	62·50	36·75	1·73
4	82·75	61·50	35·25	2
5	82·25	60·50	34·50	2·24
7	81·75	59·25	32·75	2·64
9	81·25	58·25	31·75	3
11	80·75	57·50	30·75	3·32
13	80·50	57·00	30·00	3·6
15	80·25	56·75	29·75	3·86
20	80·00	56·00	29·00	4·46
25	79·75	55·50	28·50	5
30	79·50	54·50	28·00	5·47
35	79·25	54·50	27·75	5·9
40	79·00	54·25	27·50	6·3
50	78·50	54·00	27·25	7·05
60	78·25	53·25	26·75	7·74
90	78·00	53·00	25·00	9·46
120	77·75	52·75	24·25	10·95
180	77·50	52·25	24·00	13·45
24 hrs.	70·40	45·60	18·03	37·95

$T_v = \dfrac{C_v t}{d^2}$ At $U_v = 90\%$ consolidation, then time factor $T_v = 0.848$

Using Taylor & Merchant's method to evaluate t_{90}

then $C_v = \dfrac{T_v (d^2)}{t_{90}} = \dfrac{.848 (d^2)}{t_{90}}$

where d = drainage path length = $\dfrac{\text{initial thickness}}{2} = \dfrac{18.80}{2} = \underline{9.9\text{mm}}$

with 2 way drainage.

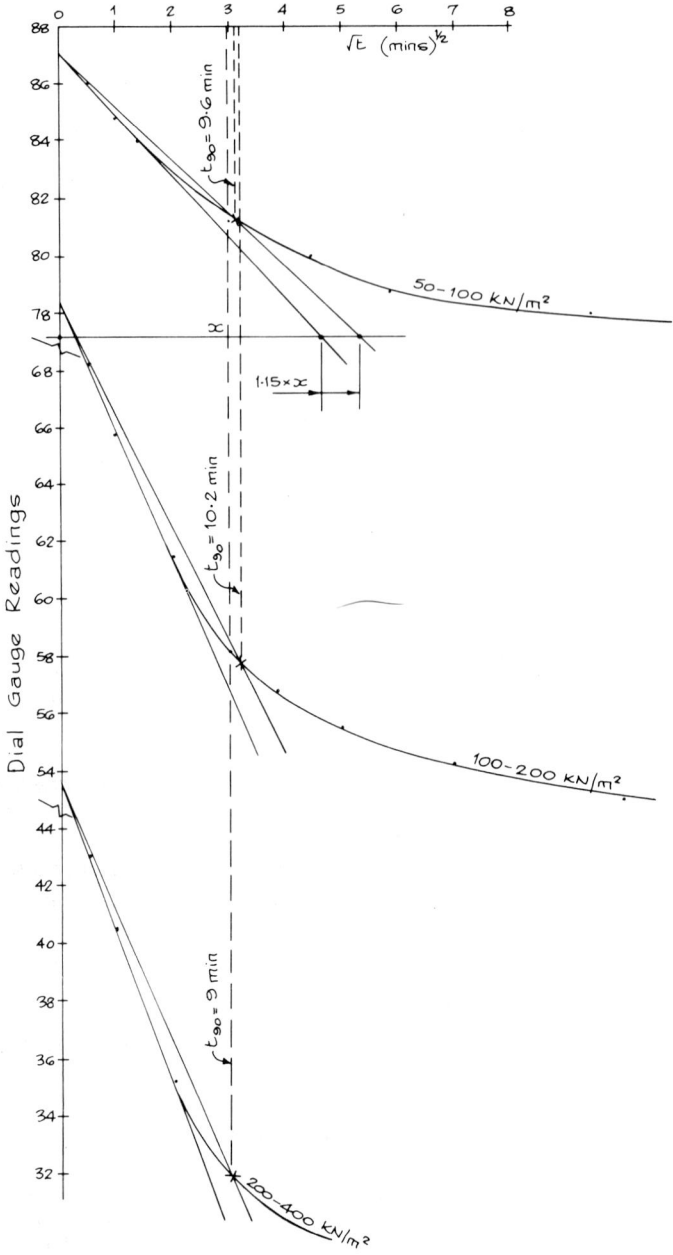

Coefficient of Consolidation (Cv)

$$C_v = \frac{T_v\,(d^2)}{t_{90}}$$

Increment 50 to 100 kN/m² $t_{90} = 9.6$ min

$d = \frac{18.8}{2} = 9.4$ mm

$$C_v = \frac{0.848\,(9.4)^2}{9.6} \times 0.5256 = \underline{4.102\ m^2/yr}$$

Increment 100 to 200 kN/m² $t_{90} = 10.2$ min

$d = \frac{18.6}{2} = 9.3$ mm

$$C_v = \frac{0.848\,(9.3)^2}{10.2} \times 0.5256 = \underline{3.779\ m^2/yr}$$

Increment 200 to 400 kN/m² $t_{90} = 9$ min

$d = \frac{18.4}{2} = 9.2$ mm

$$C_v = \frac{0.848\,(9.2)^2}{9} \times 0.5256 = \underline{4.192\ m^2/yr}$$

Coefficient of volume decrease (Mv)

$$M_v = \frac{\Delta H}{H_o}\left(\frac{1}{\Delta P}\right)$$

Increment 50 to 100 kN/m²

$H_o = 18.80$ mm

$\Delta H = \frac{90 - 70.4}{100} = 0.196$ mm

$\Delta P = 50$ kN/m²

$$M_v = \frac{.196}{18.8} \times \frac{1}{50} = 0.0002085\ mm^2/kN$$

$$M_v = 2.085 \times 10^{-4}\,mm^2/kN$$

Increment 100 to 200 kN/m²

$H_o = 18.80 - 0.196 = 18.604$ mm

$\Delta H = \frac{70.40 - 45.60}{100} = 0.248$ mm

$\Delta P = 100$ kN/m²

$$M_v = \frac{.248}{18.604} \times \frac{1}{100} = 0.000133\ mm^2/kN$$

$$M_v = 1.33 \times 10^{-4}\,mm^2/kN$$

Increment 200 to 400 kN/m²

$H_o = 18.604 - 0.248 = 18.356$ mm

$\Delta H = \frac{45.60 - 18.03}{100} = 0.2757$ mm

$\Delta P = 200$ kN/m²

$$M_v = \frac{.2757}{18.356} \times \frac{1}{200} = 0.0000751\ mm^2/kN$$

$$M_v = 0.751 \times 10^{-4}\,mm^2/kN$$

of excess pore-water pressures but no means is available for monitoring the pore-water pressures during testing.

5. The size of the sample is often too small to be considered to be representative of the mass, particularly where the clay being tested has several natural features affecting drainage rates, such as fissures or organic intrusions. The effect of trimming such a small sample can also seriously disturb the soil structure.

6. The pore-water pressure levels during testing do not usually correlate with the pore-water pressure levels at the field scale since the back-pressure usually employed in the test is atmospheric, or very nearly so.

5.3 The Rowe Consolidation Cell

In an attempt to overcome the difficulties associated with the Casagrande oedometer, particularly with respect to sample size and knowledge of pore-water pressure behaviour, Rowe and Barden [4] introduced a new consolidation cell giving many distinct advantages. The cell is available in diameters of 76 mm, 152 mm and 254 mm. More specialised testing has resulted in the use of 508 mm diameter cells.

A typical set-up for a vertical consolidation test, being undertaken on a 254 mm diameter sample, is shown in Fig. 5.13 and other relevant details are given in Fig. 3.8a (page 57).

The cell is manufactured from an aluminium bronze casting which is non-corrodible. A cell base of machined mild steel is bolted to the body and sealed by an 'O'-ring. This base is in contact with the underside of the sample and the top of the sample is covered by saturated filter paper and a saturated sintered bronze disc. A convoluted rubber jack then bears on the top of the sintered bronze disc and hydraulic pressure applied to the jack provides the total pressure applied to the top of the sample. The inside of the cell is lightly greased to overcome side-friction effects as the sample consolidates.

A hollow brass spindle is attached to the contact face of the rubber jack and passes through the aluminium cover to the cell which is bolted to the top of the cell body. This spindle allows the pore water, expelled from the soil during consolidation, to pass through the sintered bronze disc, up the centre of the spindle and, via nylon tube and a Klinger valve, to the back-pressure system. As the soil consolidates, the spindle follows the jack in contact with the top of the sintered bronze disc. A dial gauge mounted externally to the top of the spindle can thus monitor vertical settlement of the sample. This dial gauge is fitted to a rigid bridge framework bolted to the top cover of the cell.

A Klinger valve in the top cover allows entry of the water to pressurise the sample. This water can itself be pressurised from a regulated air supply feeding an air–water cylinder or, where small volume change is expected, from a constant-pressure mercury pot system.

A convenient form of back-pressure unit is to employ an elevated water-bottle or, if available headroom does not allow this, another mercury-pot

Fig. 5.13 Rowe consolidation cell, 254 mm diameter

constant pressure system. The facility then exists to provide a full range of total pressures (from the air–water cylinder) together with appropriate back-pressures (from elevated water-bottles). Thus any levels of expected total, pore-water and effective pressures can be attained to suit the particular site under consideration.

Complete monitoring of pore-water pressures is achieved by means of a sintered bronze or ceramic disc fixed centrally in the upper surface of the base of the cell and a 3 mm diameter hole drilled through the base leading to a Klinger valve and then immediately to a pressure transducer mounted in a brass block. A further Klinger valve then connects the transducer, by copper tubing, to a null indicator de-airing system and hand pump. This allows de-airing of the associated leads, repeated flushing and calibration of the transducer against a digital voltmeter prior to the setting up of a sample for testing.

5.3.1 Procedure

With undisturbed samples, the sample tube is lined up horizontally and concentric with the cell body and slowly jacked into the cell body. Care must be taken to ensure that the applied pressure during extrusion is recorded and not allowed to become excessive. The sample is carefully trimmed flush at the base of the cell with a wire or saw. De-aired water is spread over the base of the cell and the body is transported to the cell base and bolted to it. Extreme care is required to minimise the risk of entrapping air and support of the sample is required during transportation. It is obviously preferable to extrude the samples adjacent to the prepared cell bases.

After the body and base are bolted, the sample is carefully scraped to the required thickness at the top, the filter paper and sintered bronze are placed on top, the whole is flooded with de-aired water and note taken of the depth of scraping. From the physical dimensions of the cell, the thickness of sample can now easily be evaluated. The jack is then lowered into the water covering the sample, thus displacing it, and the lid and jack are securely bolted. The jack is filled with water and a small bedding-pressure from a simple elevated water-bottle is applied. This bedding-pressure is often of the order of 100 mm head of water and just allows priming of the system while draining against zero back-pressure.

The air regulator and elevated bottle are adjusted according to the values of total pressure and back-pressure that have been selected. The Klinger valve connected to the drainage lead from the spindle is closed, the dial gauge reading is noted, the pressure transducer is switched into the circuitry and the required total pressure is applied to the jack. When the output on the digital voltmeter corresponds to a pore-water pressure equal to the applied pressure, consolidation can be started. With the drain closed, if the applied jack pressure equals the pore pressure at the base of the sample then the sample is fully saturated and Terzaghi's concept of the spring and piston analogy is confirmed.

To start consolidation, note of the dial gauge and pore-water pressure (via the voltmeter) is taken. Opening the drainage valve at a specific time then clearly defines the start of consolidation. Readings of time, dial gauge and pore-water pressure are then recorded at sensible intervals and the progress of consolidation is recorded. When the pore-water pressure at the base of the sample corresponds to the back-pressure, primary consolidation can be taken as being complete (i.e. the mid-plane excess pore-water pressure has been dissipated).

The time taken to reach this stage obviously depends upon the permeability of the soil, the magnitude of the pressure increment applied and the thickness of the sample. With large samples, this period is one of days rather than hours and may well be in excess of one week, and throughout this time checks must be made to ensure that the test is proceeding in a satisfactory manner and that levels of total applied pressure and back-pressure are being constantly maintained. In addition, regular observations of time, settlement and pore-water pressure are recorded either manually or automatically.

It is good practice, after priming the sample under a small head, to bed the sample to a state of effective stress corresponding to the state existing at field scale. This re-creates conditions of void ratio in the sample similar to those in the field. Any subsequent pressure increments can then be applied from this state of *in situ* void ratio and corresponding effective pressure.

On completion of testing, the sample is subjected to immediate moisture-content and specific-gravity determinations to establish correlation between void ratio and sample thickness.

Typical results are given in Datasheet No. 15.

5.3.2 Summary

1. The testing of samples that are considered to be representative of the mass can readily be undertaken using the Rowe consolidation cell. The results of tests on large samples are far more realistic than those obtained by means of the Casagrande oedometer. This is especially true for soils that have distinct characteristics which can affect drainage rates and the deformation.

2. The start of consolidation is markedly defined by the opening of a drainage valve and time can be allowed, after applying the pressure increment, for the pore-water pressure to build up.

3. The ability to measure and monitor the pore-water pressure is particularly advantageous. Pore-water pressure dissipation rates obtained in the laboratory can then readily be compared with those at field scale by installing piezometers and monitoring field build-up and dissipation.

4. Complete automatic control of settlement and pore-water pressures can be incorporated and thus save valuable technician time.

5. Testing against *in situ* levels of pore water (as a back pressure) can easily be arranged. In addition, permeability tests can readily be performed with the same sample used for consolidation tests.

6. The hydraulic loading system can be pressurised by an air-supply line or, where build-up of load tests is required, by a motorised mercury pot system that will elevate the pressure at a required rate.

7. Similarly, swelling test studies can be undertaken so long as safeguards are provided to ensure that pore-water pressure remains essentially positive. This may necessitate adjustment of the applied total pressure and back pressure.

8. Radial consolidation characteristics can also be studied by using 'inward radial' test techniques – where drainage is to a central model sand drain – or by using 'outward radial' tests to a peripheral porous lining

Vertical Consolidation using Rowe Cell.

SETTLEMENT

Date	Time	Total Time (min)	Dial Gauge	Diff (mm)	Settlement (mm)
27 Nov.	9.00am	0	15.85	—	—
		1	15.82	.03	.03
		2	15.81	.01	.04
		3	15.81	.00	.04
		4	15.80	.01	.05
		5	15.79	.01	.06
		6	15.79	.00	.06
		7	15.78	.01	.07
		8	15.78	.00	.07
		9	15.78	.00	.07
		10	15.77	.01	.08
		12	15.76	.01	.09
		14	15.75	.01	.10
		16	15.75	.00	.10
		18	15.74	.01	.11
		20	15.73	.01	.12
		25	15.71	.02	.14
		30	15.70	.01	.15
		35	15.68	.02	.17
		40	15.67	.01	.18
		45	15.66	.01	.19
	10.00am	60	15.62	.04	.23
		95	15.56	.06	.29
	11.00am	120	15.50	.06	.35
		160	15.44	.06	.41
		255	15.31	.13	.54
	2.00pm	300	15.25	.06	.60
		343	15.21	.04	.64
	4.00pm	420	15.17	.04	.68
	6.00pm	540	15.12	.05	.73
28 Nov.	9.00am	1440	14.66	.46	1.19
29 Nov.	9.00am	2880	14.60	.06	1.25
30 Nov	9.00am	4320	14.55	.05	1.30
1 Dec.	9.00am	5760	14.53	.02	1.32
2 Dec.	9.00am	7200	14.51	.02	1.34
3 Dec.	9.00am	8640	14.50	.01	1.35
4 Dec.	9.00am	10080	14.48	.02	1.37

PORE WATER PRESSURE

Date	Time	Total Time (min)	DVM Reading (mV)	Diff. from Initial	% PWP Dissipation
27 Nov.	9.00am	0.0	235	—	—
		2	233	2	1.11
		4	233	2	1.11
		6	233	2	1.11
		8	233	2	1.11
		10	232	3	1.66
		13	229	6	3.32
		23	226	9	4.98
		33	222	13	7.18
		43	218	17	9.39
		53	218	17	9.39
		73	212	23	12.71
		93	207	28	15.47
		113	201	34	18.79
		153	193	42	23.20
		193	184	51	28.18
27 Nov.	1.03pm	243	178	57	31.49
		303	168	67	37.02
		343	164	71	39.23
		643	129	106	58.56
		943	110	125	69.06
		1243	95	140	77.35
28 Nov.	10.43am	1543	79	156	86.19
28 Nov.	3.43pm	1843	58	177	97.79
29 Nov.	1.43am	2443	57	178	98.34
29 Nov.	11.43am	3003	56	179	98.90
30 Nov.	11.43am	4443	54	181	100.00
1 Dec.	11.43am	5883	54	181	100.00

Total pressure = 180 kN/m² (FINAL)

Back pressure = 40 kN/m²

Effective pressure = 140 kN/m² (FINAL)

Total pressure = 70 kN/m² (INITIAL)

Effective pressure = 30 kN/m² (INITIAL)

INITIAL THICKNESS = 72.13 mm.

INITIAL MOISTURE CONTENT = 24.7 %

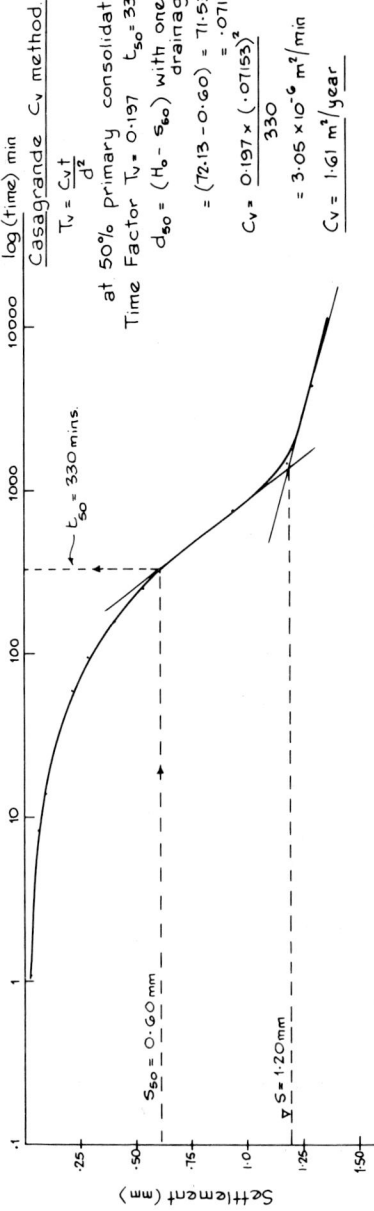

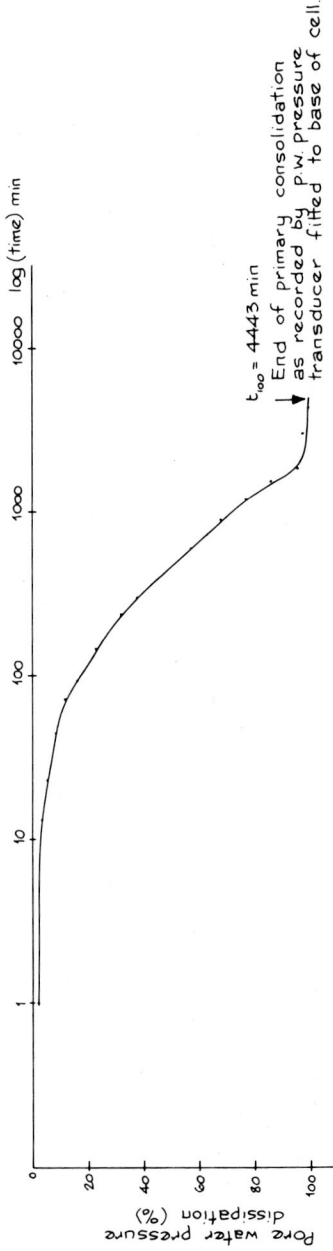

Casagrande C_v method

$$T_v = \frac{C_v t}{d^2}$$

at 50% primary consolidation
Time Factor $T_v = 0.197$ $t_{50} = 330 \text{ min}$
$d_{50} = (H_0 - s_{50})$ with one way drainage
$= (72.13 - 0.60) = 71.53 \text{ mm}$
$= .07153 \text{ m}$

$$C_v = \frac{0.197 \times (.07153)^2}{330}$$

$$= 3.05 \times 10^{-6} \text{ m}^2/\text{min}$$

$$\underline{C_v = 1.61 \text{ m}^2/\text{year}}$$

log (time) min

$t_{50} = 330 \text{ mins}$

$s_{50} = 0.60 \text{ mm}$

$\triangledown s = 1.20 \text{ mm}$

Settlement (mm)

.25 .50 .75 1.0 1.25 1.50

log (time) min

$t_{100} = 4443 \text{ min}$

End of primary consolidation
as recorded by p.w. pressure
transducer fitted to base of cell.

Pore water pressure
dissipation (%)

0 20 40 60 80 100

totally surrounding the cylindrical surface area of the sample. The latter type of test is quicker to perform from a commercial viewpoint but the samples used need to be smaller than the nominal size of cell being used.

9. Results of tests performed on large-diameter samples relate well to practice and there is strong indication that the testing of large samples, incorporating features of the soil, will become more and more common.

5.4 Determination of the Coefficient of Consolidation (C_v) by Pore-water Pressure Dissipation Tests

One distinct advantage of using the Rowe cell for the determination of the consolidation characteristics of soil is the facility to observe pore-water pressures. This facility is not usually available with conventional oedometers. Pore-water pressure measurement allows the coefficient of consolidation to be determined from 'dissipation' tests rather than by observing volume change (or settlement) and using a curve-fitting technique to settlement versus time curves.

Many arguments can be put forward to say that it is better to determine C_v in this way since consolidation behaviour is directly related to the dissipation of excess pore-water pressures set up by the application of some external load. One particular advantage of using dissipation tests is that both the commencement and termination of primary consolidation are clearly defined by pore-water pressure observations. The starting point being when the pore-water pressure is equal to the applied loading, which also incidentally verifies full saturation of the soil, and the end of primary consolidation being when the pore-water pressure value equates to the back-pressure equivalent to the *in situ* hydrostatic value. Equally any other point of percentage dissipation can be defined with equal clarity since the pore-water pressure at any instant of time within the consolidation process is observed and recorded.

5.4.1 Procedure

For instance, if it is necessary to determine the point of 50% dissipation of excess pore-water pressure, then the frequency of readings of pore-water pressure can be intensified at this time to ensure accuracy of determination of the time taken to reach t_{50}.

The following illustrates an example.

In situ effective pressure = 50 kN/m^2
Hydrostatic pressure   = 20 kN/m^2 (back-pressure)
Total *in situ* pressure   = 70 kN/m^2 (applied total pressure).

Suppose the applied load increment is 50 kN/m^2 then the experimental procedure would be as follows.

(*a*) Close drainage valve on consolidation cell.

(b) Increase applied total pressure to *in situ* value (70 kN/m²) plus applied increment (50 kN/m²) giving a total applied pressure of 120 kN/m².

(c) With drain closed, pore-water pressure will rise to 70 kN/m², being back-pressure of 20 kN/m² plus applied increment of 50 kN/m². This will be recorded, and displayed and printed by the digital voltmeter via the pressure transducer. This will also serve as a check on the state of full saturation.

(d) Define the start of consolidation by simultaneously opening the drainage valve, starting the timing device and arranging for scanning, and print-out, of the pore-water pressure readings.

(e) The point of 50% primary consolidation is clearly defined as that time when the pore-water pressure has dissipated to 50% of the applied increment. In this case the pore-water pressure value would be

$$\tfrac{1}{2}(\text{applied increment}) + (\text{back-pressure value}) = \tfrac{1}{2}(50) + (20)$$
$$= 45 \text{ kN/m}^2$$

Continual monitoring of the pore-water pressure will define when this value is being approached and an increase in the frequency of scanning on the digital voltmeter will clearly define the time when a pore-water pressure of 45 kN/m² is achieved, even if this time is outside of normal laboratory working hours.

(f) When the pore pressure measuring device records 20 kN/m², primary consolidation will be complete and the excess pore pressure, generated by the applied load, will be 100% dissipated.

(g) Graphically, the pore-water pressure dissipation versus time (to a logarithmic scale) graph can then be plotted and t_{50} clearly defined.

For the example given then a plot of the form shown in Fig. 5.14 will result. This value of t_{50} can then be used in the expression

$$T_v = \frac{C_v(t)}{d^2}$$

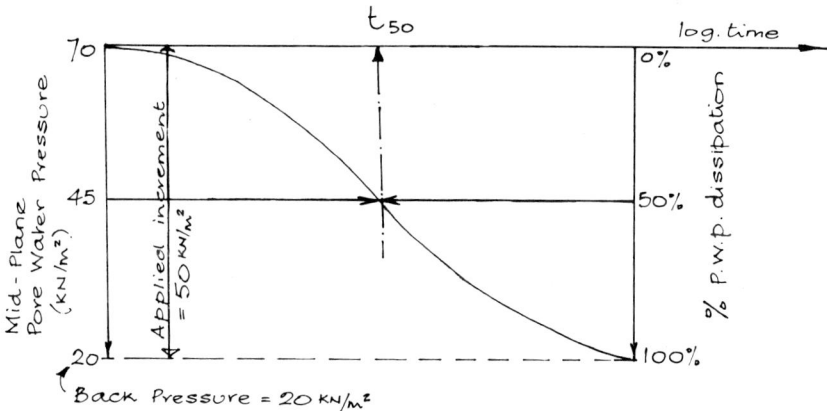

Fig. 5.14

where $T_v = 0.38$ (and *not* 0.197) = Terzaghi time factor

If $t_{50} = 1200$ min

$d = 72.4$ mm (taking account of settlement up to this time
of 1200 min)

then

$$0.38 = \frac{C_v(1200)}{(72.4)^2}$$

$$C_v = 1.66 \text{ mm}^2/\text{min}$$

The reason for $T_v = 0.38$ for dissipation tests at 50% and not 0.197 in settlement (or volume change) tests is that the excess pore-water pressure is a point value measured at the mid-plane of the sample, whilst $T_v = 0.197$ relates to an average value of degree of consolidation for the *full depth* of the sample. In diagram form, when a sample consolidates there are important differences in time factor (T_v) values, which are related to degree of consolidation (U_v) values.

5.4.2 Mid-plane Pore-water Pressure

From Fig. 5.15 it can be seen that the point values of pore-water pressure vary with distance from the drainage boundary. Almost immediately consolidation starts the layers of soil adjacent to the boundary are 100% consolidated whilst those layers farthest from the drainage boundary are, possibly, only slightly consolidated. As time progresses, and drainage continues, the excess pore-water pressure distribution with depth of sample varies as shown (Fig. 5.15).

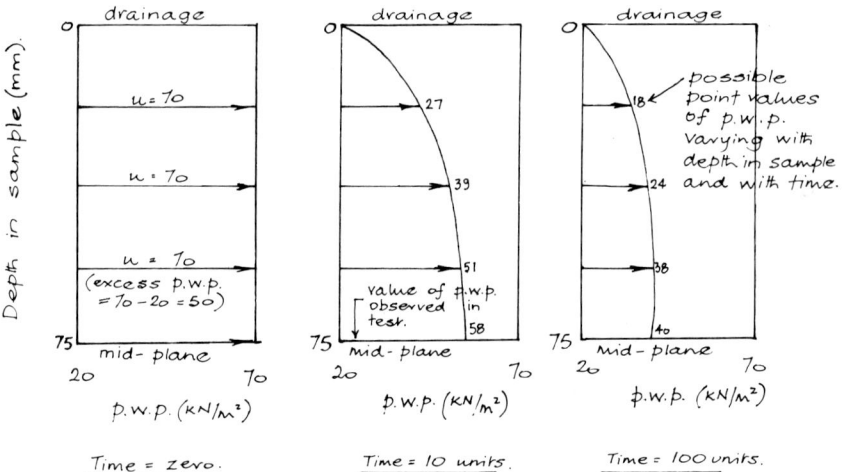

Fig. 5.15

Thus at any instant of time, there is a varying degree of consolidation throughout the layers of the soil. Consolidation is more quickly completed the nearer the element of soil is to the drainage boundary. The elements of soil farthest from the drainage medium (i.e. the mid-plane) are adjacent to the point of observation of the pore-water pressure (i.e. the base of the cell).

The U_v and T_v values given as $U_v = 90\%$ when $T_v = 0.848$ and $U_v = 50\%$ when $T_v = 0.197$ relate to *average* degrees of consolidation for the full depth of the sample. More simply, the average degree of consolidation could be 70% whilst point values could be 100% next to the drain and 42% at the mid-plane.

Thus when the mid-plane pore-water pressure has been dissipated by 50% (i.e. fallen from 70 to 45 kN/m²) the average degree of consolidation of the sample will be greater than 50%, hence the higher T_v value (Fig. 5.16).

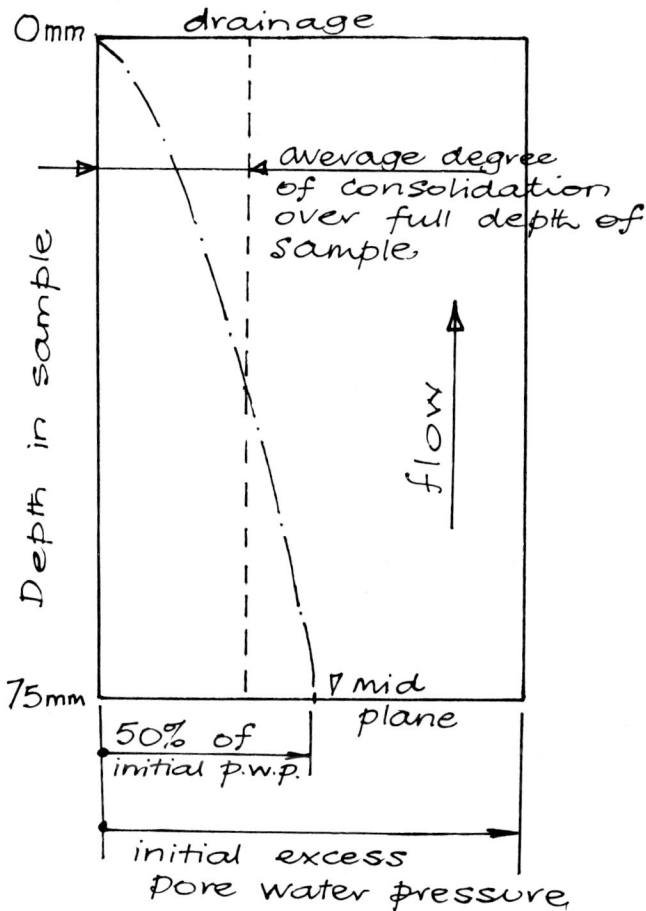

Fig. 5.16

The use of such a technique to find the coefficient of consolidation, therefore, has two major advantages, namely,

(i) There is no need to revert to a curve-fitting technique, involving graphical methods.

(ii) The time of 50% dissipation is accurately determined and recorded and there is no need to perform the test for a period longer than is necessary for, say, 60% dissipation of pore-water pressure if the object of the testing is to determine C_v values only. This can be of commercial use in shortening testing and 'turn round' times but would be valueless if the 100% dissipation time and/or C_α values are of interest.

Similar dissipation tests can be performed using a triaxial cell set-up and Bishop and Henkel[8] give full details of such test arrangements.

CHAPTER 6

Soil Compaction Tests

2.5 kg rammer method
4.5 kg rammer method
Vibrating hammer method for granular soils

6.1 Introduction

In a loose state, soil consists of solid particles, water and air. Compaction of
the soil, usually by mechanical means, reduces the air voids with the aim of

controlling subsequent moisture-content changes;
achieving a state of increased unit weight;
increasing the shear strength of the soil;
reducing the permeability;
making the soil less susceptible to settlement under load, especially
repeated loading from, say, traffic.

To achieve any of these aims when constructing dams, embankments, retaining walls, roads or runways, strict control of the construction methods is
required. Laboratory tests have been developed so that the specification for
field compaction can be written and checks can be made on the fill material
as placed and compacted to ensure that the specification is being met. If
compaction is not carried out thoroughly during construction, future loading of the material when it forms part of a permanent structure will cause
further undesirable compaction with associated deformation and likely
damage.

When water is added to a dry soil, each soil particle adsorbs a film of
water, which surrounds it. As more water is added, the film of water
increases in thickness and permits easier sliding of particles relative to each
other. This is a 'lubrication' process and the added water is replacing air

within the voids of the soil. At a certain point during the addition of more water, the added water occupies space that could be occupied by solid particles and the air content remains constant. This occurs at a high degree of saturation. Obviously there is an optimum quantity of water, for a particular soil being compacted in a particular manner, at which there is a maximum mass of solid matter per volume while a minimum quantity of air is maintained.

The function of the laboratory compaction tests is thus to determine this optimum quantity of water and the corresponding unit weight of the given soil. The tests must give results that can be compared to those obtainable at field scale and consequently the effort applied during compaction in the laboratory must be directly related to the effort applied by compaction plant on site. As compaction plant has developed over the last 40 years, both in size and in effort provided for compaction, the effort applied in standard laboratory tests has been increased to keep pace. It is of interest to note that the optimum amount of water required for good compaction is dependent upon the effort applied in compacting the soil.

In the model soil sample shown in Fig. 6.1, V_A = volume of air, V_w = volume of water and V_s = volume of solid, so that the total volume V = $V_A + V_w + V_s$.

If a is the percentage air voids, defined as

$$\frac{a}{100} = \frac{V_A}{V}$$

or

$$V_A = \frac{aV}{100}$$

then

$$V = \frac{aV}{100} + V_w + V_s$$

or

$$V(1 - \frac{a}{100}) = V_w + V_s \tag{6.1}$$

If the density of water is ρ_w, specific gravity of the solid particles is G_s and the moisture content of the soil is m,

$$m = \frac{V_w \rho_w}{V_s G_s \rho_w} = \frac{V_w}{G_s V_s}$$

or

$$V_w = mG_s V_s$$

Substituting in eq. (6.1) for V_w,

$$V(1 - \frac{a}{100}) = mG_s V_s + V_s = V_s(1 + mG_s)$$

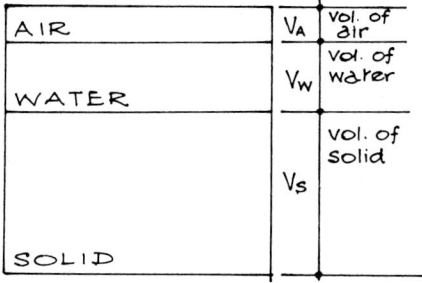

Fig. 6.1

so that

$$V = \frac{V_s(1 + mG_s)}{1 - (a/100)}$$

The dry density ρ_d of the soil is by definition

$$\rho_d = \frac{V_s G_s \rho_w}{V} = \frac{V_s G_s \rho_w}{V_s(1 + mG_s)}[1 - (a/100)]$$

that is

$$\rho_d = \frac{G_s \rho_w[1 - (a/100)]}{(1 + mG_s)}$$

Note: In applying the above equation, a is expressed as a percentage, e.g. $a = 8\%$, while m is not expressed as a percentage, i.e. if $m = 18\%$ then use $m = 0.18$.

For zero air-void content ($a = 0\%$),

$$\rho_d = \frac{G_s \rho_w}{1 + mG_s} = \text{saturation dry density}$$

The relationship between ρ_d and m obtained from laboratory compaction tests is usually of the form shown in Fig. 6.2.

The effect of varying the compactive effort is shown in Fig. 6.3. It is seen that the general effect of increased effort is to reduce the optimum moisture content and increase the maximum dry density.

To allow valid comparisons to be made between various soil types, standard laboratory compaction tests have been laid down in BS 1377: 1975 as follows:

2.5 kg rammer method
4.5 kg rammer method
Vibrating hammer method for granular soils.

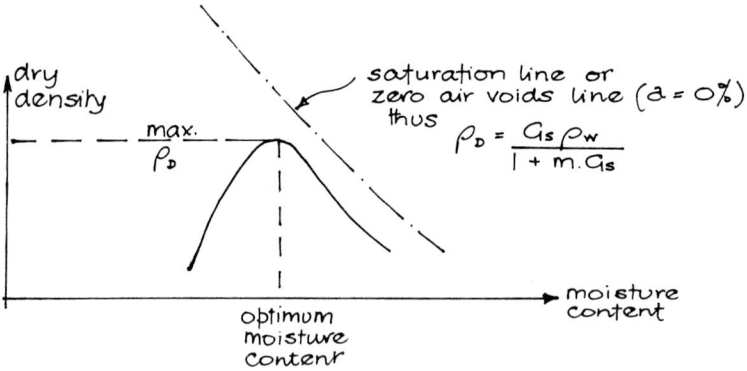

saturation line or zero air voids line $(a = 0\%)$

thus $\rho_D = \dfrac{G_s \rho_w}{1 + m.G_s}$

Fig. 6.2

Each test involves the determination of the relationship between the dry density and the moisture content of a soil. The tests have been developed from work done by Proctor in 1933 and adopted by the American Association of State Highways Officials. Consequently the original tests were often referred to as the Proctor test or the AASHO test.

6.2   2.5 kg Rammer Method

In the test the relationship between the dry density and moisture content is determined by compacting the soil with a 2·5 kg rammer falling freely through a 300 mm height. The rammer has a circular end area of 50 mm diameter and compacts the soil in a mould of 105 mm diameter, internal effective height 115·5 mm and volume 1000 c.c. The base of the compaction

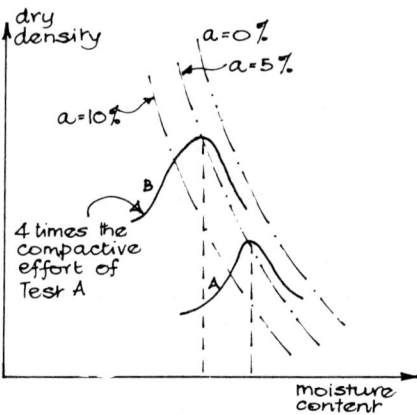

Fig. 6.3 Note that maximum dry density occurs approximately on the 5% voids line

mould is removable and rests on a solid base which usually takes the form of a steel base-plate cast into a concrete block. A detachable collar, 50 mm high, is also attached to the mould.

The device for ensuring that the drop of the rammer is maintained at 300 mm above the soil surface is either mechanical or hand-operated.

Five kg of air-dried soil which passes the 20 mm BS test sieve is thoroughly mixed with a known amount of water. Generally the initial moisture content after mixing should be 5% for coarse-grained soils; a moisture content 10% *below* the plastic limit of fine-grained soils is acceptable. Care must be taken to ensure that mixing is thorough and that an even distribution of water throughout the sample is achieved. This may well mean storing the sample in a sealed container, after mixing, for up to about 20 hours when dealing with certain highly plastic soils.

The soil is then compacted into the mould which has been previously weighed (M_1). The extension collar is attached during compaction and the soil is compacted in three layers with equal quantities of soil per layer. Each layer is given 27 blows of the free-falling rammer and the mould is slightly rotated after each blow to ensure that the rammer makes contact with the full surface area of soil during compaction.

After completion of compaction the soil level should be no more than about 6 mm above the level of the top of the mould with the extension collar removed; this 6 mm or so of soil is then carefully scribed level and the soil plus mould is weighed (M_2).

The compacted soil is then quickly removed from the mould and a sample is taken for moisture-content m determination.

$$\text{Bulk density} = \frac{M_2 - M_1}{1000} \, (\text{Mg/m}^3)$$

where M_1 = mass of mould (g)
M_2 = mass of mould and compacted soil (g)
1000 = volume of mould (cm^3).

Since

$$\rho_d = \rho/(1 + m) = \text{dry density (Mg/m}^3)$$

where m is the moisture content of the soil, therefore

$$\rho_d = \frac{M_2 - M_1}{1000(1 + m)}$$

where ρ_d = dry density (Mg/m^3). (Note that, if $m = 6\%$ is the measured compacted moisture content, then $m = 0.06$ is used in the above expression.)

The soil is then broken up, mixed with the remaining soil, passed through the 20 mm BS sieve and mixed with more water to give a higher moisture content. It is usually found convenient to use moisture-content increments of $1\frac{1}{2}\%$ for coarse-grained soils and 3% for fine-grained soils.

The whole procedure is repeated at least five times at various moisture contents, and corresponding dry density and moisture contents are evaluated. A graph of ρ_d versus m is plotted and the appropriate maximum value of dry density and related optimum moisture content are noted. It is obviously important to ensure that the curve of ρ_d versus m show a definite peak value and consequently the range of moisture content values chosen must be such that this peak value is clearly evident in the results.

6.3   4.5 kg Rammer Method

The procedure for this method corresponds to that used in the 2·5 kg rammer method, with the following important differences.

The rammer used weighs 4·5 kg and has a free fall of 450 mm. In addition, the soil is compacted in *five* layers with twenty-seven blows per layer. Every other aspect of the test is the same.

Typical results for both types of test are given in Datasheet No. 16.

6.4 Vibrating Hammer Method

In this test the relationship between dry density and moisture content is investigated, usually for granular soils passing the 37·5 mm BS test sieve. The mould used for compaction is different from that of two previous tests, being a California Bearing Ratio mould, 152 mm diameter by 127 mm deep, plus a 50 mm detachable collar.

The means of compaction is an electrical vibrating hammer with a tamper attached. The tamper has a diameter of 145 mm at the soil-contact face and should not exceed 3 kg in mass. The power consumption of the hammer should be 600–750 W and the operational frequency should be in the range 25–45 Hz. BS 1377: 1975 details, in Note 2 to test 14, a test procedure to check whether a hammer complying with the test requirements is in satisfactory working order. In this important secondary test a sample of Leighton Buzzard silica sand is used and the ability of the hammer to achieve a minimum specified dry density is tested. Readers are referred to BS 1377: 1975 for comprehensive details.

6.4.1 Procedure

A 25 kg sample of air-dried soil passing the 37·5 mm BS test sieve is thoroughly mixed with a quantity of water to give a specified moisture content, usually about 4%.

The soil is compacted in the C.B.R. mould of known mass M_1 in three

Datasheet Nº 16

Compaction Test.

		Test Nº	1	2	3	4	5	6	7	8	9		
Bulk Density Determination	Mass of Mould + Wet Soil	(g)	6818	6821	6895	6970	7022	7052	7062	7039	7002	M_2	
	Mass of Mould	(g)	5139	5139	5139	5139	5139	5139	5139	5139	5139	M_1	
	Mass of Wet Soil	(g)	1679	1682	1756	1831	1883	1913	1923	1900	1863	M_2-M_1	
	Bulk Density, ρ (Mg/m²)		1.679	1.682	1.756	1.831	1.883	1.913	1.923	1.900	1.863	$\dfrac{M_2-M_1}{1000}$	
Moisture Content Determination	Container Nº		31	33	45	336	9	4	14	37	30		
	Wet Soil + Container	(g)	16.1	13.4	19.8	14.9	20.8	14.3	27.8	30.5	21.2		
	Dry Soil + Container	(g)	15.7	12.9	18.6	13.9	18.8	13.0	24.3	26.2	18.4		
	Container	(g)	8.4	5.0	4.9	5.3	4.8	5.3	5.0	4.7	4.7		
	Dry Soil	(g)	7.3	7.9	13.7	8.6	14.0	7.7	19.3	21.5	13.7		
	Water	(g)	0.4	0.5	1.2	1.0	2.0	1.3	3.5	4.3	2.8		
	Moisture Content	(%)	5.48	6.33	8.76	11.63	14.29	16.88	18.13	20.00	20.44		
	$1+m$		1.0548	1.0633	1.0876	1.1163	1.1429	1.1688	1.1813	1.2000	1.2044		
	Dry Density, ρ_D = $\rho/1+m$ (Mg/m³)		1.563	1.582	1.615	1.640	1.648	1.639	1.626	1.583	1.547		

G_s = 2.65

4.5 kg Rammer method:
Blows per layer = 27
Nº of layers = 5
Fall of rammer = 450 mm

Max. Dry Density = 1.649 Mg/m³

Optimum Moisture Content = 14.4 %

COMPACTION TEST

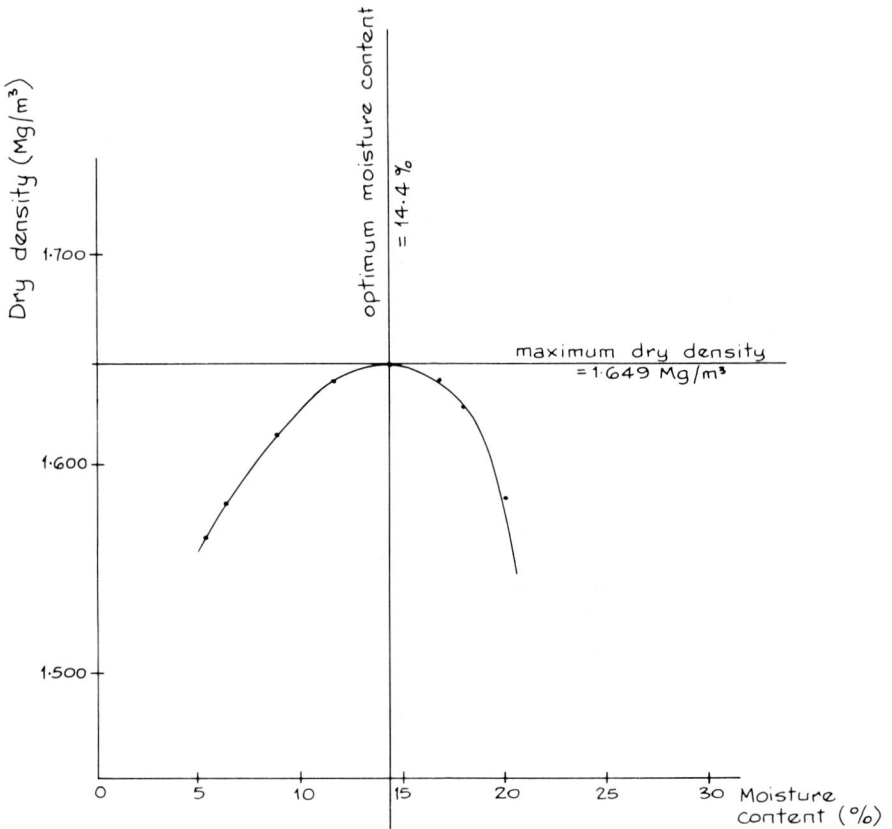

equal layers, such that the final compacted height is 130 mm $\pm$ 3 mm. Each layer is subject to vibration from the hammer for 1 minute and during this time a constant downward force is applied to the hammer. This force must be between 300 and 400 N and this value includes the mass of both the tamper and the hammer. The function of the downward force is to make sure that the necessary degree of compaction is achieved and to prevent bouncing of the hammer on the soil surface. Obviously this operator-applied pressure requires experience and 'dummy' runs have to be performed. Note 6 to Test 14 in BS 1377: 1975 suggests a method of testing that the applied pressure is correct.

After compaction of the soil has been successfully completed, loose material on the surface of the sample is removed and observations of depth to the nearest 0·5 mm are taken from a straight-edge spanning across the collar to the mould. Four such depths are recorded, evenly distributed over the surface area of the sample and at least 15 mm from the sides of the C.B.R. mould. If the computed thickness of sample is outside the limits of 127 mm and 133 mm, the result must be rejected and the test carried out again. If the thickness of the sample is acceptable, the mould, complete with collar and compacted soil, is weighed (M_2); the soil is then removed and a representative sample taken for moisture-content determination (m).

The whole procedure is repeated at least five times in increments of increasing moisture content of 1·5% until a plot of ρ_d versus m, with appropriate peak values, is obtained.

$$\rho = \text{Bulk density} = \frac{M_2 - M_1}{181\cdot5 \times d} \quad (\mathrm{Mg/m^3})$$

where $181\cdot5 = \pi/4 \times (152 \text{ mm})^2 \times (1/100) = $ cross-sectional area of sample (cm^2)

$d = $ mean thickness of compacted sample (cm)

$M_2 = $ mass of mould, collar and compacted soil (g)

$M_1 = $ mass of mould and collar (g).

Therefore

$$\rho_\mathrm{d} = \text{dry density} = \rho/(1 + m) \quad (\mathrm{Mg/m^3})$$

Figure 6.4 shows details of the moulds used together with a suitable type of tamper for the vibratory test.

6.4.2. Summary

1. The aim of compaction tests is to relate laboratory work to field compaction achievement. The 2·5 kg rammer method, especially for highly permeable gravels and sands, provides only an approximate indication of the state of compaction that can be achieved at full scale. This is mainly due to the lack of comparison between laboratory and field compactive efforts.

The test results, therefore, tend to indicate higher optimum moisture contents and lower maximum dry densities than can be achieved on site with modern compaction equipment. Similar reservations apply to the testing of the same highly permeable gravels and sands with the 4·5 kg hammer. Consequently, the vibrating hammer test has been developed for such soils and with its use the correlation between laboratory and field states of compaction is much improved: generally, optimum moisture contents are similar, while laboratory maximum dry density values are only slightly in excess of those which can be achieved on site.

2. Where soils contain a significant quantity of larger stones, boulders or rock fragments, the test result is considered to be valid if the percentage of particles greater than 20 mm diameter which is removed before testing is less than 5% by weight of the whole sample. If, however, the percentage of large particles is greater than 5%, BS 1377: 1975 (Note 2 to test 13) should be consulted as to how this problem can be tackled.

3. It is obvious that, for important projects involving large quantities of compacted soil, field trials are economically and practically advantageous. In such trials a variety of combinations of the number of passes, types of plant and layer thicknesses can be investigated. A combination is then chosen which will achieve the specified state of compaction in the most economical way.

4. A very important aspect of compaction work is the control and supervision on site. This usually involves carrying out regular field checks of bulk density and moisture content. Various *in situ* methods are detailed in BS 1377: 1975. In addition, it is important to maintain checks on the quality of the imported fill, methods of compaction and layer thickness. In the field, the weather plays an important part in compaction; compaction work must not proceed when there is excessive rainfall, frost or, indeed, prolonged dry spells. The aim of such control is to meet the requirements of the specification for the contract.

5. In recent years there has been an increase in the number of specifications for compaction work in which an acceptable percentage of air voids is required to be achieved at field scale. This seems logical when one remembers that the aim of compaction is to expel air from the voids and thus generate the desirable soil properties. Some specifications may well require various parts of the site to be compacted to different extents, dependent upon the function of the compacted fill. In addition, the limit of acceptable moisture content must also be specified.

6. The main function of a laboratory compaction test programme may be to classify soils into categories according to suitability for use as fill material in embankment and fill construction, so that the suitability of a particular soil may be assessed for use as fill at a particular site.

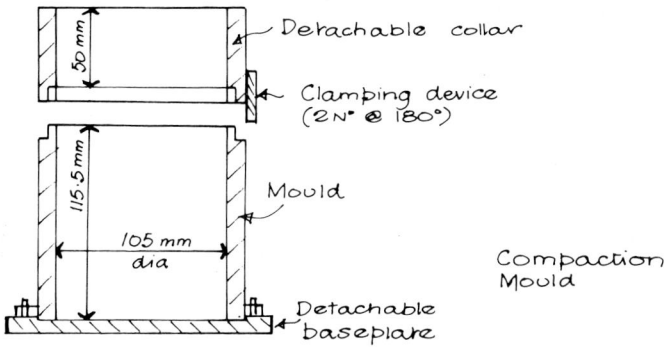

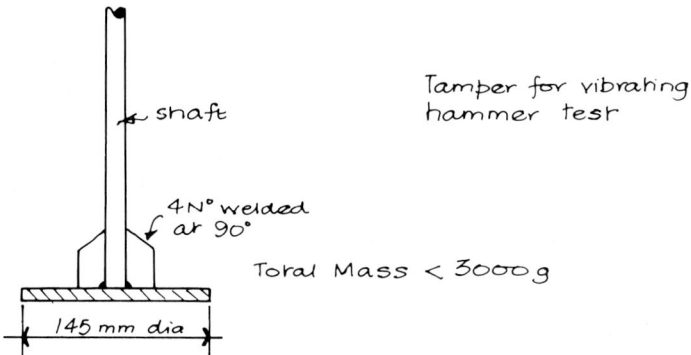

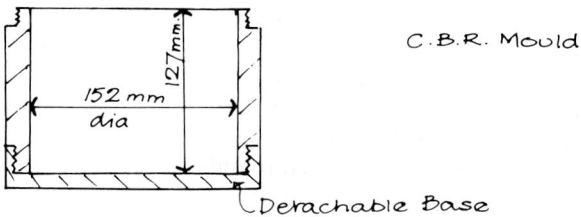

Fig. 6.4(a) Compaction mould
  (b) Tamper for vibrating hammer test
  (c) C.B.R. mould

In Situ Testing: The Standard Penetration Test

Introduction

The Standard Penetration Test (S.P.T.) is an *in situ* test performed, usually, whilst carrying out a site investigation. Since the development of the test it has become widely accepted as part of standard procedure in Soil Mechanics. Any technician or engineer involved in laboratory work, whether testing samples or analysing test results, will invariably meet 'N' values, listed as part of a site investigation report.

Much published work is available on the relation of the results of standard penetration tests to important soil properties. Attempts have been made to relate 'N' values to shear strength parameters, bearing capacity, soil density, state of compaction, as well as relating 'N' values to the performance of structures.

The test remains as possibly the most important *in situ* test and is particularly useful in sand or gravel deposits.

It is vital that the test is performed exactly as specified in BS 1377: 1975, since minor deviations from the standard practice have been shown to affect predictions made about the properties of the deposit.

One added benefit is that during the test, with most soils, some form of disturbed sample can be obtained and this sample will be useful for identification and descriptive purposes. However, with coarser soils or rocks, such a sample is not available since the penetration involves a solid cone rather than an open-ended split sampling device.

The major advantages of the test are that it is reasonably simple to perform and it is very useful in soils which are difficult to sample (e.g. filled ground) or where the quality of sample may be inferior. In addition it is relatively cheap in equipment and labour costs when carried out as a part of a conventional site investigation and will readily act as a 'check' on subsequent laboratory values of soil properties obtained from undisturbed samples.

Outline of Procedure

During the site investigation it is often specified that standard penetration tests will be performed in granular layers at 1·5 m centres down the borehole. The borehole, which has previously been sampled up to the point of the test, is carefully cleaned out and the standard penetration equipment is fitted to the drilling rig. This comprises essentially a 'split-spoon' sampler (see Fig. A1.1) 457 mm long and 50 mm outside diameter, a driving assembly and guide arrangement.

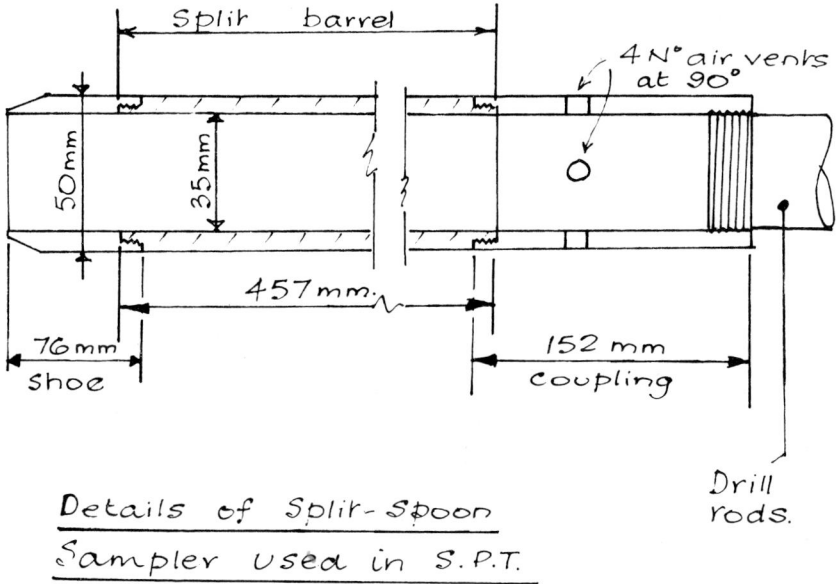

Details of Split-Spoon Sampler used in S.P.T.

Fig. A1.1

A hammer and anvil arrangement is fitted so that a 65 kg mass is able to fall freely a distance 760 mm to enable the sampler to penetrate the base of the borehole (see Fig. A1.2). Initially the sampler is driven to 150 mm below the base of the borehole to avoid any base soil disturbed by creating the borehole. The number of blows required to drive the sampler a further 300 mm beyond this depth is noted.

It is good practice to record the number of blows for the 'initial' 150 mm penetration and then to record the number of blows for the 'further' 300 mm penetration separately. This latter value is then taken as the 'N' value for the soil.

If less than 300 mm penetration is achieved then the drilling record must state both penetration achieved and the number of blows required to reach this penetration.

It is important to ensure that

(i)   The sampler is thoroughly clean before each S.P. test.

Note:-
Water level
kept at least
up to ground
level

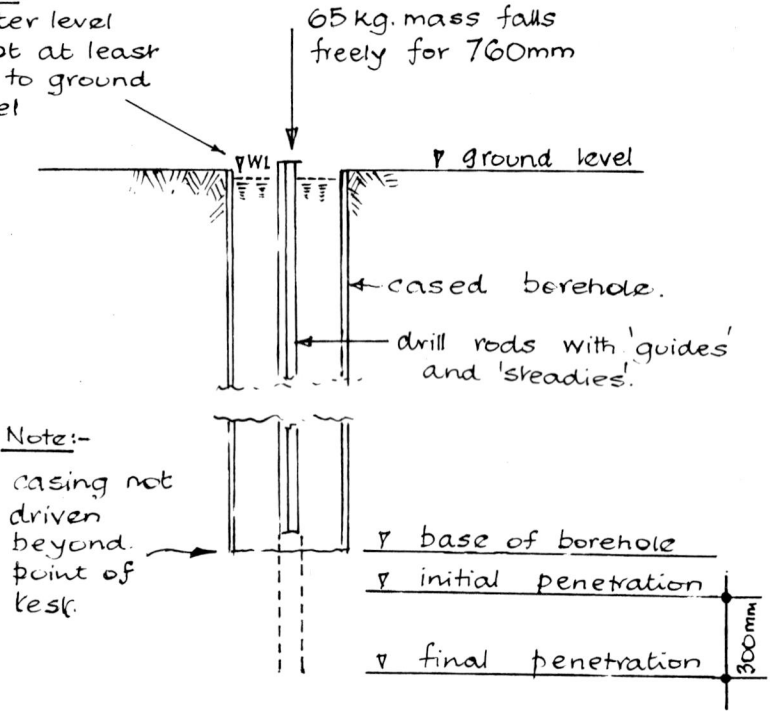

65 kg. mass falls
freely for 760mm

▽ ground level

▽WL

casing not
driven
beyond.
point of
test.

cased borehole.

drill rods with 'guides'
and 'steadies'.

Note:-

▽ base of borehole

▽ initial penetration

▽ final penetration

300mm

Fig. A1.2

(ii) There is minimum friction between the guides and the falling hammer and at the operating winch, i.e. free fall of the hammer must be maintained.

(iii) The rods are of stiffness specified in BS 1377: 1975.

(iv) Upward flow of water into the base of the borehole is avoided if at all possible. Such flow would tend to loosen the soil at the base of the borehole.

With fine or silty sands, that are saturated, experience has shown that an erroneous value of the relative density can be obtained. This relative density is usually higher than the realistic value. In such cases where the recorded value of 'N' is greater than 15, it is common practice to apply an empirical correction, namely:

$$N' = 15 + \tfrac{1}{2}(N - 15)$$

Obviously throughout the depth of a borehole, with continual values of 'N' being obtained, there will be a variation of 'N' value with depth and it is useful to plot this relationship. Some engineering judgement will then be required to decide how the variation in 'N' value with depth is to be interpreted. Often an average 'N' value is obtained for each borehole but such a choice, and indeed, the choice of a value from borehole to borehole will depend upon the particular problem being investigated.

Soil Properties

Note: It is unwise to infer the values of probable soil properties, but from a student viewpoint it is useful to be aware of probable order of magnitude values so that as a test proceeds the student can be conscious of any errors that may be developing.

Useful ranges of values of various properties are given in the following tables for guidance only.

Permeability

Table A2.1

Permeability (metres/second)	Soil type
$K = 1$ to about 10^{-2}	Gravel
$K = $ about 10^{-3} to 10^{-4}	Sand
$K = 10^{-5}$ to about 10^{-9}	Fine sands and fine sand clay mixtures
$K = 10^{-10}$ to about 10^{-11}	Clays

As mentioned in the text, a clay with sand lenses, organic intrusions or fissures may well be more permeable than the above values. Similarly the permeability of a soil varies with the effective pressure and void ratio. A commonly accepted relationship is that the void ratio (e) is linearly related to the logarithm of the coefficient of permeability (K). Thus a soil with a wide range of void ratio values will undoubtedly have wide-ranging permeability values.

In addition, the vertical coefficient of permeability value will normally be lower than the coefficient of permeability in a horizontal direction.

Undrained Cohesion Values Related to Soil Description

From a conventional undrained triaxial text on a saturated clay soil or an unconfined compression test, the resulting cohesion value is often used as a qualitative means of describing the soil as shown in Table A2.2.

Table A2.2

C_u = undrained cohesion (kN/m²)	Description
Over 150	Very stiff or hard
75–150	Stiff
40–75	Firm
20–40	Soft
Less than 20	Very soft

The above descriptive terms are often further qualified as shown in Table A2.3.

Table A2.3

C_u = undrained cohesion (kN/m²)	Description	
40–50	40 to 75 = Firm	Soft to firm
50–75		Firm
75–100	75 to 150 = Stiff	Firm to stiff
100–150		Stiff

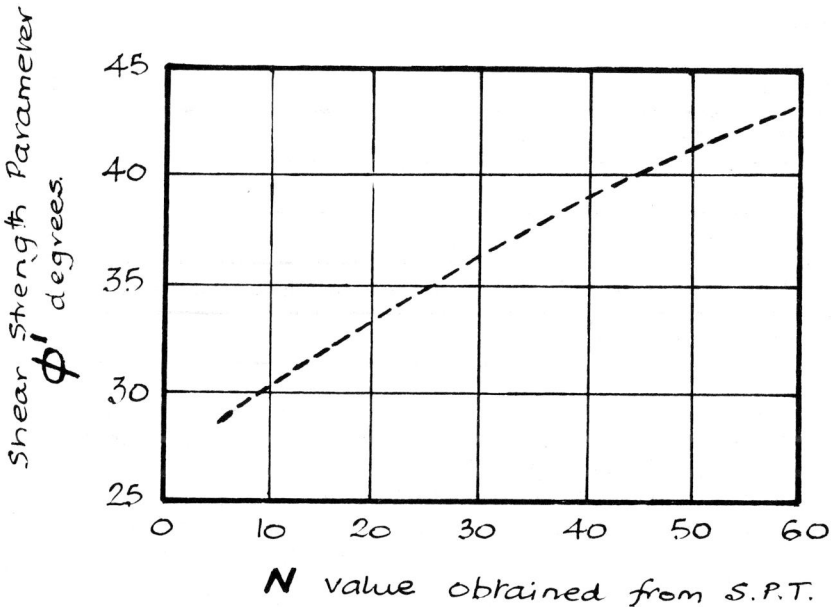

Fig. A2.1 (After Peck, Hanson and Thornburn)

Relative Density of Sands and Gravels Related to Standard Penetration Test '*N*' Values

(N = number of blows for 300 mm penetration)

Table A2.4

'*N*' value	Relative density
0 to 4	Very loose
4 to 10	Loose
10 to 30	Medium dense
30 to 50	Dense
Over 50	Very dense

Consolidation Characteristics

Since consolidation characteristics are very much a function of the effective pressure of a soil, it is important to ensure that values obtained for, say, m_v = coefficient of volume change or C_v = coefficient of compressibility, correspond to the range of pressures anticipated at the full-scale.

As a *guide*, the range of m_v values expected for soils of UK origin is given in Table A2.5.

Table A2.5

Coefficient of volume decrease m_v (m²/kN) – range of values	Soil type
Less than 0·00005	Heavily overconsolidated soil
0·00005 to 0·0001	Boulder clay
0·0001 to 0·0003	London clay
0·0003 to 0·0015	Normally consolidated clay

In addition it is possible to infer coefficient of consolidation values (C_v) by using the expression defining C_v in the classical Terzaghi consolidation theory, viz:

$$C_v = \frac{K}{\gamma_w m_v}$$

where K = coefficient of permeability
γ_w = unit weight of water
m_v = coefficient of volume decrease.

References

1. British Standard 1377: 1975 *Methods of Test for Soil for Civil Engineering Purposes*. British Standards Institution, London.
2. Casagrande, A. (1948) Classification and identification of soils. *Transactions of the American Society of Civil Engineers*, **113**, 901.
3. Wagner, A. A. (1957) The use of the unified soil classification system by the Bureau of Reclamation. *Proceedings 4th International Conference Soil Mechanics and Foundation Engineering*, Vol. 1. ISSMFE, London.
4. Rowe, P. W. and Barden, L. (1966) A new consolidation cell. *Geotechnique*, **16** (2), 162–170.
5. Wilkinson, W. B. (1968) *Permeability and Consolidation Measurements in Clay Soils*. PhD Thesis, University of Manchester.
6. Bishop, A. W., Garga, C. E., Andresen, A. and Brown, J. D. (1971) A new ring shear apparatus and its application to the measurement of residual strength. *Geotechnique*, **21**, 273–328.
7. Rowe, P. W. and Barden, L. (1964) The importance of free ends in triaxial testing. *Journal of the Soil Mechanics and Foundations Division, ASCE*, **90** (SM1), 1–27.
8. Bishop, A. W. and Henkel, D. J. (1962) *The Measurement of Soil Properties in the Triaxial Test*. Edward Arnold, London
9. Henkel, D. J. and Gilbert, G. D. (1952) The effect of the rubber membrane on the measured triaxial compression strength of clay samples. *Geotechnique*, **3**, 20.
10. Casagrande, A. (1936) The determination of the pre-consolidation load and its practical significance. *Proceedings 1st International Conference Soil Mechanics and Foundation Engineering*, Vol. 3. ISSMFE, Cambridge.
11. Schmertmann, J. H. (1953) Estimating the true consolidation behaviour of clay from laboratory test results. *Proceedings of the American Society of Civil Engineers*, **79**, Separate No. 311.
12. Skempton, A. W. (1954) The pore pressure coefficients A and B. *Geotechnique*, **13** (4).

Index